Quantum Mechanics

Embarking on a journey into the realm of quantum mechanics can be a daunting task for anyone. Its puzzling mathematics and bewildering predictions often leave individuals feeling confused and disheartened. But what if there was a different approach — one aimed to cultivate an understanding of quantum mechanics from its very foundations? This is the ambition of this book. Rather than treating quantum mechanics as an inception, the author takes a Socratic perspective, tracing the genesis of its key ideas back to the well-established roots of classical mechanics. The works of Lagrange, Hamilton, and Poisson become guiding lights, illuminating the path towards comprehension. Through a colloquial yet pedagogical narrative, the book delves into the elements of classical mechanics, building a solid framework of familiarity that paves the way for comprehending quantum mechanics.

Designed as a companion for undergraduates undertaking quantum mechanics modules in physics or chemistry, this book serves as an invaluable support. It equips learners with the essential knowledge necessary to grasp the foundations of quantum mechanics. As such, it proves equally beneficial for MSc and PhD scholars, and post-doctoral researchers. Its colloquial tone captivates the curiosity of any reader eager to delve into the mysteries of this enthralling field.

Giuseppe Pileio is currently an Associate Professor of Physical Chemistry at the School of Chemistry of the University of Southampton, UK, where he teaches Physical Chemistry, Quantum Mechanics, Molecular Spectroscopy and Nuclear Magnetic Resonance. At the University of Southampton, Giuseppe also run a research group that specializes in developing theory, simulations, and methodology in the field of Nuclear Magnetic Resonance. Giuseppe is co-author of more than 70 scientific papers published in peer-reviewed journals, and has already published a University Textbook on the dynamics of nuclear spins.

Quantum Mechanics

Inception or Conception? A Classical Preamble to Quantum Mechanics

Giuseppe Pileio

CRC Press
Taylor & Francis Group
Boca Raton London New York

CRC Press is an imprint of the
Taylor & Francis Group, an **informa** business

First edition published 2024
by CRC Press
2385 NW Executive Center Drive, Suite 320, Boca Raton FL 33431

and by CRC Press
4 Park Square, Milton Park, Abingdon, Oxon, OX14 4RN

CRC Press is an imprint of Taylor & Francis Group, LLC

ISBN: 9781032584300 (hbk)
ISBN: 9781032605883 (pbk)
ISBN: 9781003459781 (ebk)

DOI: 10.1201/9781003459781

Typeset in Minion
by Deanta Global Publishing Services, Chennai, India

To my daughter Ludovica, may this serve as a reminder to consistently seek the logical progression of ideas and concepts

Contents

Acknowledgements

Writing a book is a solitary endeavour; yet, it arises from countless interactions: with books, educators, colleagues, students, and introspection. The inspiration for this book struck me when I realized that the students I taught quantum mechanics to regarded it as a collection of perplexing equations unrelated to the Chemistry they were studying. While I strongly disagreed with their perception of relevance, I empathized with their sense of bewilderment, as it mirrored my own experiences as a student two decades ago.

In truth, as I will elucidate in the Introduction, I believe that the way the subject is taught — particularly to undergraduate Chemistry students, based on my humble experience — is to blame for such sentiments. It is precisely this concern that I aim to address within these pages. Thus, my foremost and deepest gratitude goes to all my students at the University of Southampton, particularly Calypso Burger, Lauren Cheney, Cat O'Rourke, Martina Rabrenovic, Andrew Smith, and Kim Yeap, who attended the lecture series upon which this book is founded. Their evident enthusiasm for the subject, insightful comments, and thought-provoking questions not only helped me streamline the discussion to its core aspects but also guided me in choosing an appropriate language and level of understanding.

I owe a great debt to my long-time friend and now esteemed colleague, Salvatore Mamone, for meticulously reviewing the draft and providing numerous invaluable comments. However, I must stress that I alone am accountable for any errors or omissions in the final version. I extend my sincere gratitude to the anonymous reviewers of my book proposal, all of whom expressed kind words of appreciation for this endeavour.

Lastly, I would like to express my heartfelt appreciation to the team at CRC Press — Taylor & Francis for accepting this book for publication and for their unwavering support throughout the process, ensuring that this vision becomes a reality.

CHAPTER 1

Introduction

QUANTUM MECHANICS IS UNDENIABLY a challenging subject, both for lecturers to teach and for students to understand. A few years ago, I conducted a poll among my second-year chemistry students asking whether they find quantum mechanics difficult, and if yes, why. Although that was certainly not a statistically meaningful sample of students studying this subject, most answers gravitated around *it is a very difficult subject* and this is mostly because of *its too complex mathematical description.*

In my humble experience and from my modest point of view, mathematics itself is not the primary obstacle to comprehension. The mathematical foundations of quantum mechanics, at least the portion pertinent to chemistry undergraduates (to whom I have the privilege of teaching), are relatively straightforward. It is worth noting that while the formalism may be new to students, many problems can be solved, demonstrated, and understood without necessitating the arduous solution of complex mathematical equations. Nonetheless, the fact that second-year chemistry students perceive the mathematics of quantum mechanics as overly intricate is a concerning signal that should not be disregarded, but such conversation warrants separate consideration.

While acknowledging the mathematical difficulties associated with quantum mechanics, it is essential to recognize other significant factors contributing to the challenges faced by both students and instructors. Of particular importance, in my opinion, is the *students' relative unfamiliarity with the subject*, which they often encounter relatively late in their studies. Additionally, the *seemingly peculiar predictions* of quantum theory and, not insignificantly, the *chronological* (and, one might argue, *illogical*)

DOI: 10.1201/9781003459781-1

order in which these concepts are typically taught and learned further compound those difficulties.

The issue of sequencing has long perplexed me, both as a student and later as a lecturer on the subject. I was relieved to discover that this concern was not unique to me but a topic of discussion among physicists. Leonard Susskind aptly articulated this conundrum as a "quantum chicken and egg" dilemma: which should come first, classical or quantum mechanics?[1]

My personal dilemma closely aligns with Susskind's observations, but it centres more specifically on the genesis of the concept of quantum mechanics. I ponder whether quantum mechanics emerged as an organic outgrowth of classical mechanics, or if it was introduced as an external notion — a question of *inception* versus *conception*. I find the contrasting nature of these two terms captivating, as their meanings carry an air of elusiveness. Inception denotes a literal *beginning*, while conception represents *the initial step toward creation*. Furthermore, conception serves as a synonym for *idea*, but the phrase *conception of an idea* emphasizes the act of formulating an idea within one's own mind, distinct from inception, where the idea has been implanted by an external source.

To the best of current human understanding, quantum mechanics is more fundamental than classical mechanics, the latter approximating quantum mechanics very well when dealing with entities of macroscopic sizes. The laws of classical mechanics were, however, discovered first, as often happen in science where models are refined by successive approximations. Moreover, being particularly good to describe macroscopic phenomena, the *approximate* laws of classical mechanics become familiar to us very earlier in our studies. Many of those laws are part of the science curriculum in primary and secondary schools across the world. Hence, another dilemma: what to teach first, quantum mechanics that is *logically more fundamental* than classical mechanics but harder to comprehend or classical mechanics that is only an approximation but whose laws are much easier to understand? Personally, I find myself oscillating between both sides of the argument, recognizing the advantages and disadvantages of both approaches. However, if compelled to decide, I lean towards teaching classical mechanics first, followed by quantum mechanics. Yet, it is precisely this pedagogical approach that presents a problem. Classical mechanics is not included in the curriculum for non-physics undergraduate students, nor it is typically integrated as a preamble to quantum mechanics courses. Consequently, it becomes evident how the introduction of new and unfamiliar concepts of quantum mechanics can lead to issues such as stress, frustration, and a lack of understanding among students. In my view, the primary struggle stems from the inability to establish connections between these novel concepts and prior knowledge. As I expressed in my textbook on spin dynamics, I firmly believe that "we truly comprehend things only when they are contextualized within frameworks that our minds are already acquainted with".

In this book, my intention is to explore the essential readings that students, as well as any curious reader, should engage with before delving into the realm of quantum mechanics. By emphasizing the word *must* in italics, I meant to highlight the benefits that can be derived from such preparatory reading. Furthermore, the italicization of the word *read* underscores the significance of engaging with the material, even if complete comprehension is not immediately achieved. This is a book to be *read*, but not necessarily *understood*. The purpose of this book is to establish familiar frameworks that students can relate to when approaching the subject of quantum mechanics. I acknowledge that the phrase *a book to be read, but not necessarily understood* may raise some controversy, and I shall make an effort to clarify its intended meaning. Drawing upon Martin Broadwell's four-stage learning theory, I propose that the primary goal of this book is to assist readers in transitioning from stage 1, which is *unconscious incompetence*, to stage 2, which is *conscious incompetence*. To achieve this, attentive reading should be sufficient.

Nonetheless, I aspire to present the material in a manner that facilitates comprehension, enabling readers to progressively advance to stage 3, characterized by *conscious competence* in learning. For a detailed discussion on the content encapsulated within these stages of learning, I encourage interested readers to explore and study Martin Broadwell's original work.

Before delving into the unravelling of the subject within this book, it is essential for me to elucidate the motivations behind its creation. With a *laurea magistralis* and a PhD in chemistry, a discipline focused on the study of atoms and molecules, I have come to recognize the pervasive presence of quantum mechanics in all facets of chemistry. Whether it be theoretical chemistry, computational chemistry, molecular spectroscopy, or nuclear magnetic resonance, quantum mechanics permeates every level. Moreover, the fundamental comprehension of chemical reactions employed in organic, inorganic, and biochemical synthesis relies heavily on the principles of quantum mechanics. In my current role as a University Associate Professor, I have had the opportunity to teach various courses in physical chemistry, with a significant emphasis on quantum mechanics, molecular spectroscopy, and nuclear spin dynamics. Consequently, I have acquired a certain level of knowledge on the subject and, more importantly, I have faced the challenge of effectively teaching it to undergraduate chemistry students in a manner that benefits them. It was during the organization of teaching materials for these courses that the *teaching order dilemma* first struck me. Over the years of instructing the subject, I have gathered valuable insights and, crucially, gained perspective from the students themselves. Although I attended two physics courses during my undergraduate studies, including one on classical mechanics as part of my chemistry degree, I was unaware of many of the principles that I will discuss later in this book. It was only in recent years that I became acquainted with these laws, and I couldn't help but think: *if only I had known this earlier, my understanding of quantum mechanics would have been far more profound*!

By introducing and exploring these laws within the pages of this book, my intention is to evoke the same sense of discovery and realization in my readers. I hope to ignite within them the same sentiments I experienced upon uncovering these principles.

The approach I have chosen to present the material begins with an introduction to the fundamental laws of quantum mechanics. To accomplish this, I will follow the common approach found in most quantum mechanics books, where the foundational elements of quantum mechanics are introduced as *postulates*. It is important to note that this approach contradicts what I previously mentioned, which advocated for starting with classical mechanics before transitioning to quantum mechanics. In the realm of logic, a proposition is referred to as a postulate when it is assumed as a foundational principle for a theory without requiring proof. This manner of encountering unproven statements has always intrigued me. Although it is not uncommon and widely accepted in science, it personally leaves me with a sense of something missing or not entirely comprehended. Nevertheless, once the postulates of quantum mechanics have been introduced, I will proceed to discuss the laws of classical mechanics in their various formulations presented by Newton, Lagrange, Hamilton, and Poisson. Among these formulations, Hamilton's and Poisson's representations align more closely with the *familiar framework* that will assist readers in rationalizing quantum mechanics more easily.

Subsequently, I will revisit the topic of quantum mechanics, capitalizing on the familiarity gained through this process of acclimatization. In a final appendix, I will provide a brief discussion on the concepts of ignorance, the scientific method, and two fundamental tools of logic. I believe that these topics should be included in the repertoire of every science student.

To summarize and further clarify, this book primarily targets students who are either embarking on or planning to pursue a scientific degree that encompasses courses on quantum mechanics. It emerges from my notes for the undergraduate zero-credit module titled: *Informal discussions on the fundamental laws of Nature.* The book does not aspire to be a comprehensive guide to either quantum mechanics or classical mechanics. Instead, its purpose is to serve as a *classical preamble to quantum mechanics*, providing students with a preliminary understanding of the fundamental concepts that form the foundation of this intricate field of physics. By establishing familiarity with these essential concepts, the book aims to facilitate the study of quantum mechanics, recognizing its inherent complexity. It does not seek to encompass the breadth and depth of the subject, but rather serves as a precursor that prepares students for the exploration of quantum mechanics.

NOTE

1. Quantum Mechanics, The Theoretical Minimum, by Leonard Susskind and Art Friedman, Penguin, 2014, ISBN:978-0-141-97781-2

CHAPTER 2

Quantum Mechanics (Inception)

I WOULD LIKE TO START this chapter with a note of warning that echoes the last sentence in the Introduction: in here, I am not going to present an exhaustive description of quantum mechanics, mainly because this is not the aim of the book. Rather, I shall only present the basic postulates and main results of quantum mechanics and in particular the part related to the description of the motion of an *object*. The "size" of such an object is fundamentally important for whether it is feasible to describe it with classical rather than quantum mechanics. Essentially, quantum mechanics is *necessary* when describing the mechanics of microscopic objects (particle, atoms, molecules) while classical mechanics applies (read: *it is a good and convenient approximation to quantum mechanics*) to objects of macroscopic sizes (all the way to astronomical dimensions).

The word *mechanics* stands for the part of physics that describes the motion of objects. But what does it mean to *describe*? It means to have a way to know where an object is, at any time; not just now, but also where it will be in the future and, indeed, where it has been in the past. You may immediately recognize that such knowledge would require at least two basic elements: something describing the position of the object as a function of time, and a rule to predict how such position changes with time. In addition, we often need to know where the object is (or has been) at least at one time. In many quantum mechanics courses and books, those very basic elements are often introduced in terms of postulates, i.e.

DOI: 10.1201/9781003459781-2

in terms of propositions introduced without proof. Note, though, that the fact that postulates are not proved does not mean they are unreliable; this is because postulates are based on, and consistent with, experimental observations.

In no particular order, the first postulate of quantum mechanics I present to you is about the description of the object's position; it states:

1. All properties of an object (not just its spatial position) are contained in a *wavefunction*.

The wavefunction is generically denoted as $\Psi(\vec{r},t)$ where $\vec{r}$ represents spatial coordinates and t stands for time. Thus, for a single object, $\vec{r}$ is the vector with components $\{x,y,z\}$ and x, y and z are the three axes of the Cartesian space. I know from experience that the term *wavefunction* scares many students. The simpler term *function* would surely be more tolerated, not least because $\Psi(\vec{r},t)$ is most literally a mathematical function of the variable $\vec{r}$ and t (and possibly other variables). However, we need to understand that a particle such as an electron, whose description falls within the remits of quantum mechanics, is not to be thought as an object occupying a well-defined point in space (point is here intended in its mathematical meaning, i.e., an exact location in space with no length, no width, no depth) but rather a wave (hence wavefunction) diffused over a certain volume of space. There you can begin to see some of the *peculiarity* I was talking about in the Introduction. And there is immediately more to it. As will be clearer later, some of the actual wavefunctions that describe particles such as electrons, photons or whatever microscopic objects, are imaginary (i.e. they contain i, the mathematical imaginary unit) and therefore it is hard for us to make sense of what they could represent, not least because we use to cope with real things in everyday life. Because of this fact, very early in the development of quantum theory, it became crucial to attribute a more meaningful sense to the wavefunction, a sense that we can all understand. And, once again, this was done in the form of a postulate that states:

2. The square modulus of the wavefunction is a probability density (or amplitude). This probability density can be integrated over a volume

element ($d\vec{r}$) to determine the total probability of finding the object described by such wavefunction within the specified volume element. In mathematics, this is written as:

$$Probability = \int_V \Psi^*(\vec{r},t)\,\Psi(\vec{r},t)\,d\vec{r} \qquad (2.1)$$

To make the point, the description of nature offered by quantum mechanics is based on the concept of wavefunction, the mathematical element that contains all information about an object. Incidentally, the asterisk used next to the wavefunction symbol in $\Psi^*(\vec{r},t)$ indicates the complex conjugate of the wavefunction, obtained by flipping the sign of each imaginary unit that the wavefunction $\Psi(\vec{r},t)$ may contain. Being generally imaginary, though, the wavefunction does not carry an immediately evident meaning. Its square modulus, however, represents the probability density to find the object in some (to be defined) location in space. Note that for what is said, we seem to be able to know where the object is only within a certain probability, just to add something more to the list of *oddities* you are going to encounter here.

But, how does quantum mechanics propose to extract information from a wavefunction? The answer to this is yet again expressed in the form of postulates. A first postulate deals with physical observables, i.e., with all quantities that are measurable in a laboratory, such as energy, position, angular momentum, and so on. It states:

3. Physical observables are represented by mathematical *operators* that act on the wavefunction to extract the value of the physical entity they represent.

Operators are usually indicated with a hat over the symbol, as in $\hat{O}$. A second postulate is introduced to deal with the prediction of what would result from *a large number of* experimental measurements of such observables, and states:

4. The result of a large number of experimental measurements of the observable represented by the operator $\hat{O}$ is predicted by the following calculation:

$$\langle \hat{O} \rangle = \frac{\int_V \Psi^*(\vec{r},t)\hat{O}\Psi(\vec{r},t)d\vec{r}}{\int_V \Psi^*(\vec{r},t)\Psi(\vec{r},t)d\vec{r}} \tag{2.2}$$

The quantity above, indicated as $\langle \hat{O} \rangle$, takes the formal name of *expectation value* of the operator $\hat{O}$ or, better, the *expectation value* of the physical observable represented by the operator. If the operator is the one that extracts the linear momentum of the object, then the calculation in Eq. (2.2) returns the expectation value of the linear momentum.

Why calling it *expectation value* and not just *value*? To answer this question, I first need to discuss how *operators* act on *wavefunctions*. Most generally, an operator turns a wavefunction into another wavefunction; in mathematics this is written as:

$$\hat{O}\Psi(\vec{r},t) = \phi(\vec{r},t) \tag{2.3}$$

Occasionally, but not rarely at all, the result is:

EIGENEQUATION

$$\hat{O}\psi(\vec{r},t) = \lambda\psi(\vec{r},t) \tag{2.4}$$

with λ being a generally complex number. In the last equality, we read that the effect of the operator acting on a wavefunction may result in leaving the wavefunction unaltered — other than multiplied by a complex number! Equations of such kind are called *eigenequations*. In an eigenequation, the complex numbers λ are called the *eigenvalues* of the operator $\hat{O}$, and the wavefunctions $\psi(\vec{r},t)$ are called its *eigenfunctions*. Note that I have used a different symbol to distinguish an eigenfunction, $\psi(\vec{r},t)$, from a more general wavefunction, $\Psi(\vec{r},t)$, that may or may not be an eigenfunction of that operator. Eigenequations are very important in quantum theory, especially regarding observables and expectation values. Suppose the operator is the one representing the kinetic energy of the object, Eq. (2.2) could then be used to predict the result of a large number of experimental measurements of the kinetic energy of such object to be:

$$
\begin{aligned}
\langle \hat{\text{O}} \rangle &= \frac{\int_V \Psi^*(\vec{r},t)\hat{\text{O}}\Psi(\vec{r},t)d\vec{r}}{\int_V \Psi^*(\vec{r},t)\Psi(\vec{r},t)d\vec{r}} \\
&= \frac{\int_V \psi^*(\vec{r},t)\lambda\psi(\vec{r},t)d\vec{r}}{\int_V \psi^*(\vec{r},t)\psi(\vec{r},t)d\vec{r}} \\
&= \lambda\frac{\int_V \psi^*(\vec{r},t)\psi(\vec{r},t)d\vec{r}}{\int_V \psi^*(\vec{r},t)\psi(\vec{r},t)d\vec{r}} \\
&= \lambda
\end{aligned} \tag{2.5}
$$

The result in Eq. (2.5) can be generalized in the following statement: when the wavefunction (i.e. the mathematical function representing the object) is an eigenfunction of a given operator, the result of a large number of experimental measurements of the quantity represented by the operator (in this example the kinetic energy of the object) is given by the eigenvalue of the operator.

Generally, an operator has a set of eigenfunctions, each associated with a certain eigenvalue, so if we use the subscript k to indicate the k-th wavefunction, Eq. (2.4) is generalized into:

$$
\hat{\text{O}}\psi_k(\vec{r},t) = \lambda_k\psi_k(\vec{r},t) \tag{2.6}
$$

Incidentally, giving Eq. (2.6), the following eigenequations holds:

$$
\psi_k^*(\vec{r},t)\hat{\text{O}}^\dagger = \lambda_k^*\psi_k^*(\vec{r},t) \tag{2.7}
$$

The new term in the equation above, $\hat{\text{O}}^\dagger$, indicates the *adjoint* of the operator. The adjoint of an operator acts to the left on the complex conjugate of the wavefunction to return the complex conjugate of the eigenvalue. In systems of finite dimensions, operators can be represented by matrices and the adjoint becomes the operation consisting in taking the complex conjugate of each element and then transposing (swap rows with columns) the whole matrix. Eigenfunctions associated with distinct eigenvalues are *orthogonal* to each other. The orthogonality of two wavefunctions is expressed by the following equation:

ORTHOGONALITY OF TWO WAVEFUNCTIONS

$$\int_V \Psi_j^*(\vec{r},t)\Psi_k(\vec{r},t)d\vec{r} = \begin{cases} 1 \text{ for } j=k \\ 0 \text{ for } j \neq k \end{cases} \tag{2.8}$$

The fact that the integral of the square modulus of a wavefunction gives 1 can be articulated through the sentence: *the wavefunction that satisfies* Eq. (2.8) *is normalized to unity*. The fact that the product of two different wavefunctions integrates to zero can be articulated through the sentence: *two distinct wavefunctions that satisfy* Eq. (2.8) *are orthogonal to each other.*

I shall now consider the case in which the wavefunction representing the object is not an eigenfunction of the operator of interest, let me then mark this fact by indicating such wavefunction with a different symbol as in $\phi(\vec{r},t)$. Clearly, in this case, Eq. (2.2) cannot give the same result obtained in Eq. (2.5) since it is not possible to use an eigenequation (2.4) anymore. The solution to this problem is not trivial and requires looking at wavefunctions and operators in more depth. I hope that, by now, we understand each other when I say that operators describe observable quantities and have a set of eigenfunctions and associated eigenvalues (as in Eqs. (2.6) and (2.7)) and that wavefunctions represent all that matters about the object of interest, may this be an elementary particle, an atom, a molecule, *etc*. It is not difficult to consider that in some circumstances, say in a certain point in space or in time, or in the presence of a field of some kind, or while interacting with other objects, and in many other situations, that the object is represented by a wavefunction that happens to be not an eigenfunction of the operator representing the observable we are interested in. In such cases we need to invoke the fact that the set of eigenfunctions of an Hermitian (see below for definition) operator constitutes what in mathematics is known as a *complete and orthogonal basis set*, i.e., a set of functions using which it is always possible to write any other possible function belonging to that same space. I imagine that for many of you there is a fair amount of new and challenging mathematics in my last enunciated statement but, rather than tackling this formally, I prefer explaining it with a simple similitude that I believe can make you understand the concept quicker. Think about the more familiar 3D Cartesian (or Euclidean) space. Notoriously, in that space one can draw three orthogonal (perpendicular to each other) axes, ordinarily indicated as x, y and z.

Moreover, one can consider three vectors of unitary length, each pointing along one of those axes, and call them the *versors* of the Cartesian space, ordinarily indicated as $\vec{e}_x, \vec{e}_y$ and $\vec{e}_z$. Any point in the Cartesian space can be imagined as a vector extending from the origin of the Cartesian axis reference frame to the location in space where the point lies, let me indicate such a vector as $\vec{P}$. This latter can always be written in terms of how many multiples of each of the versors $\vec{e}_x, \vec{e}_y$ and $\vec{e}_z$ it contains. In mathematics, we would call this a linear combination (or a linear expansion of the vector in the fundamental versors of the space) and write it as:

$$\vec{P} = a\vec{e}_x + b\vec{e}_y + c\vec{e}_z \tag{2.9}$$

where a, b and c, are called *expansion coefficients*, and are, in general, real numbers. The versors $\vec{e}_x, \vec{e}_y$ and $\vec{e}_z$ can therefore be thought as forming a complete and orthogonal basis set, in the sense that any vector in the Cartesian space can be written as a linear combination of those three *versors* (a.k.a. *basis vectors*). I do not need to show that these versors are orthogonal as I believe you all know this quite well. The set is complete when there are as many basis vectors as the dimensions of the space—Cartesian space has three dimensions and therefore three orthogonal basis vectors are needed. Having understood that, now consider that, in total similitude to the versors of a Cartesian space, one can find a complete set of orthogonal functions (call them basis functions of the space or even basis vectors if you like) that can act as a basis to express any generic function of that space as a linear combination of these basis vectors. The space spanned by wavefunctions is called Hilbert space, but contrarily to the Cartesian space which has always three dimensions, the dimensions of the Hilbert space vary case by case and can even become infinite. Note also that the expansion coefficients in the Hilbert space are, generally, complex numbers and the expansion is formally written as:

WAVEFUNCTION EXPANSION

$$\phi(\vec{r},t) = \sum_{k=1}^{N} c_k \psi_k(\vec{r},t) \tag{2.10}$$

for the generic wavefunction $\phi(\vec{r},t)$ and:

$$\phi^*(\vec{r},t)=\sum_{k=1}^{N}c_k^*\psi_k^*(\vec{r},t) \tag{2.11}$$

for its complex conjugate. To make the argument as clear as possible, in this discussion $\phi(\vec{r},t)$ always indicates a wavefunction that happens to be <u>not</u> one of the eigenfunctions of the operator of interest, whereas $\psi_k(\vec{r},t)$ is one of the N eigenfunctions of such operator. c_k is the k-th complex expansion coefficient that is multiplied by $\psi_k(\vec{r},t)$ to determine how much of that particular eigenfunction is contained in $\phi(\vec{r},t)$. For completeness, it can be demonstrated that the sum of the square modulus of all expansion coefficients is the unity, i.e.:

$$\sum_{k=1}^{N}c_k^*c_k=1 \tag{2.12}$$

I am now ready to show how to predict the result of a large number of experimental measurements of the observable represented by the operator $\hat{O}$ for an object represented by the wavefunction $\phi(\vec{r},t)$ that is not an eigenfunction $\hat{O}$. In doing so, I start from Eq. (2.2) and make use the expansions in Eqs. (2.10) and (2.11) to obtain:

$$\begin{aligned}\langle\hat{O}\rangle&=\frac{\int_V\phi^*(\vec{r},t)\hat{O}\phi(\vec{r},t)d\vec{r}}{\int_V\phi^*(\vec{r},t)\phi(\vec{r},t)d\vec{r}}\\&=\frac{\int_V\sum_{j=1}^{N}c_j^*\psi_j^*(\vec{r},t)\hat{O}\sum_{k=1}^{N}c_k\psi_k(\vec{r},t)d\vec{r}}{\int_V\sum_{j=1}^{N}c_j^*\psi_j^*(\vec{r},t)\sum_{k=1}^{N}c_k\psi_k(\vec{r},t)d\vec{r}}\\&=\frac{\int_V\sum_{j=1}^{N}c_j^*\psi_j^*(\vec{r},t)\sum_{k=1}^{N}c_k\lambda_k\psi_k(\vec{r},t)d\vec{r}}{\int_V\sum_{j=1}^{N}c_j^*\psi_j^*(\vec{r},t)\sum_{k=1}^{N}c_k\psi_k(\vec{r},t)d\vec{r}}\\&=\sum_{k=1}^{N}c_k^*c_k\lambda_k\end{aligned} \tag{2.13}$$

Note that, in order to go from the second to the third line, I have used the fact that $\psi_k(\vec{r},t)$ are eigenfunctions of $\hat{O}$ with eigenvalue λ_k, i.e.:

$$\hat{O}\sum_{k=1}^{N} c_k\psi_k(\vec{r},t) = \sum_{k=1}^{N} c_k\hat{O}\psi_k(\vec{r},t)$$
$$= \sum_{k=1}^{N} c_k\lambda_k\psi_k(\vec{r},t) \quad (2.14)$$

Furthermore, to go from the third to the fourth line, I have used the fact that the basis functions are orthogonal to each other (see Eq. (2.8)).

The results in Eqs. (2.5) and (2.13) are truly remarkable. They say that if one wants to predict the result of a large number of measurements of a property represented by a given operator, in some circumstances, namely when the object is in a state that can be described by a wavefunction that is the eigenfunction of the operator, the result coincides with the eigenvalue of the operator associated to such wavefunction. Conversely, when the state object is in a state that cannot be described by a wavefunction that is an eigenfunction of the operator, then the result is the sum of all possible eigenstates of the operator weighted by the square modulus of the expansion coefficient that determine how much of each individual eigenfunction is contained in the expansion of the actual wavefunction.

If an operator has, say, only two eigenstates (and therefore just two eigenvalues) then quantum mechanics predicts that each time we measure the quantity described by the operator we will get either one or the other eigenvalue, simply because there is no other existing value! No matter if the object is described by a wavefunction that is or is not an eigenstate of the operator. Which one of the two eigenvalues will be recorded? Either one of the two eigenvalues can be recorded in every single measurement but each with a probability given by the square modulus of the coefficient in front of the associated eigenfunction in the wavefunction expansion. Thus, if the measurement is repeated a large number of times, then a weighted average is obtained (as found in Eq. (2.13)). That *large number of times* statement can therefore be more appropriately interpreted as a *statistically meaningful number of times*. Clearly, when the object is described by a wavefunction that is an eigenfunction of the operator, at each measurement one gets the same identical eigenvalue. To come back to why *expectation value* and not just *value*: the term *expectation* is chosen to highlight the probabilistic character of the prediction of experimental measurements in quantum mechanics.

The final postulate I am going to discuss, perhaps the most important for the topic of this book, concerns the way to calculate the time evolution of the wavefunction, i.e., how to predict where the object will be in the future or was in the past; it states:

5. The wavefunction evolves in time according to the following equation:

$$\frac{d}{dt}\Psi(\vec{r},t) = -\frac{i}{\hbar}\hat{H}\Psi(\vec{r},t) \tag{2.15}$$

Equation (2.15) is the notorious time-dependent Schrödinger equation you may have already heard of. As ever, the postulate introduces some new quantities: the quantity indicated as $\hbar$ is a fundamental constant of the universe and has the value of 1.054×10^{-34} J s; $\hat{H}$ is clearly an operator and, specifically, the one associated with the total energy of the object, this is best known as the Hamiltonian operator as will be clearer below. The time-dependent Schrödinger equation states that the function describing the object changes in time as the result of the effect of the energy-associated Hamiltonian operator. It would be easy to prove, but out of the scope of this book, that in the case the Hamiltonian operator is time-independent (as we will see later this essentially means that the potential energy is time-independent) then:

$$\Psi(\vec{r},t) = e^{-\frac{i}{\hbar}Et}\Psi(\vec{r}) \tag{2.16}$$

implying that the wavefunction (in fact its spatial-only part $\Psi(\vec{r})$) oscillates at the frequency $\omega = (E/\hbar)$ between real and imaginary amplitudes.

The fact that the wavefunction oscillates with time may raise concerns about the time behaviour of the probability density represented by the square modulus of the wavefunction. However, those concerns are easily addressed by showing that:

$$\begin{aligned}\Psi^*(\vec{r},t)\Psi(\vec{r},t) &= e^{\frac{i}{\hbar}Et}\Psi^*(\vec{r})e^{-\frac{i}{\hbar}Et}\Psi(\vec{r}) \\ &= \Psi^*(\vec{r})\Psi(\vec{r})\end{aligned} \tag{2.17}$$

which demonstrates how the (spatial) probability density remains constant despite the amplitude of the wavefunction oscillates with time.

Of crucial importance to the use of quantum mechanics, as well as to the narrative of this book, is the time evolution of observables and their expectation values. Thus, let me use the postulates above to derive an equation for the time evolution of the expectation value of the physical property represented by the generic operator $\hat{\mathrm{O}}$. In mathematics, time evolution (and indeed the change with respect to any variable) is expressed through a *derivative*, hence we are after the quantity $\frac{d}{dt}\langle\hat{\mathrm{O}}\rangle$. The expectation value whose time derivative we are looking for, has been introduced in postulate 4 as:

$$\frac{d}{dt}\langle\hat{\mathrm{O}}\rangle = \frac{d}{dt}\frac{\int_V \Psi^*(\vec{r},t)\hat{\mathrm{O}}\Psi(\vec{r},t)d\vec{r}}{\int_V \Psi^*(\vec{r},t)\Psi(\vec{r},t)d\vec{r}} \tag{2.18}$$

If the operator is time-independent and the wavefunction is conveniently normalized (see Eq. (2.8)), i.e., multiplied by the constant that ensures:

$$\int_V \Psi^*(\vec{r},t)\Psi(\vec{r},t)d\vec{r} = 1 \tag{2.19}$$

then Eq. (2.18) reduces to:

$$\frac{d}{dt}\langle\hat{\mathrm{O}}\rangle = \int_V \left(\frac{d}{dt}\Psi^*(\vec{r},t)\right)\hat{\mathrm{O}}\Psi(\vec{r},t)d\vec{r} + \int_V \Psi^*(\vec{r},t)\hat{\mathrm{O}}\left(\frac{d}{dt}\Psi(\vec{r},t)\right)d\vec{r} \tag{2.20}$$

One can now recall postulate 5 and combine it with Eq. (2.15) to write:

$$\begin{aligned}
\frac{d}{dt}\langle\hat{\mathrm{O}}\rangle &= \int_V \left(\frac{i}{\hbar}\Psi^*(\vec{r},t)\hat{\mathrm{H}}^\dagger\right)\hat{\mathrm{O}}\Psi(\vec{r},t)d\vec{r} + \int_V \Psi^*(\vec{r},t)\hat{\mathrm{O}}\left(-\frac{i}{\hbar}\hat{\mathrm{H}}\Psi(\vec{r},t)\right)d\vec{r} \\
&= \frac{i}{\hbar}\left(\int_V \Psi^*(\vec{r},t)\hat{\mathrm{H}}^\dagger\hat{\mathrm{O}}\Psi(\vec{r},t)d\vec{r} - \int_V \Psi^*(\vec{r},t)\hat{\mathrm{O}}\hat{\mathrm{H}}\Psi(\vec{r},t)d\vec{r}\right) \\
&= \frac{i}{\hbar}\int_V \Psi^*(\vec{r},t)\left(\hat{\mathrm{H}}^\dagger\hat{\mathrm{O}} - \hat{\mathrm{O}}\hat{\mathrm{H}}\right)\Psi(\vec{r},t)d\vec{r} \\
&= \frac{i}{\hbar}\langle[\hat{\mathrm{H}},\hat{\mathrm{O}}]\rangle \\
&= -\frac{i}{\hbar}\langle[\hat{\mathrm{O}},\hat{\mathrm{H}}]\rangle
\end{aligned} \tag{2.21}$$

(for the sign change operated across the last two lines see Eq. (2.27) below). The notation in square bracket is a mathematical operation known as the *commutator*, namely:

COMMUTATOR

$$\left[\hat{H},\hat{O}\right]=\hat{H}\hat{O}-\hat{O}\hat{H} \tag{2.22}$$

Moreover, you may have noted that I have dropped the *adjoint* (dag) symbol when moving across line three and four in the mathematical derivation of Eq. (2.21). This is because it can be easily demonstrated that the Hamiltonian, as well as <u>all other</u> operators representing physical observables, satisfies:

HERMITIANITY OF THE HAMILTONIAN

$$\hat{H}^{\dagger}=\hat{H} \tag{2.23}$$

a property called *Hermitianity.* Hermitian operators are fundamental in quantum mechanics, not least because *the eigenvalues of Hermitian operators are always real numbers.* This is an important and necessary feature: the eigenvalues of operators representing physical observables appear in the expectation value and this latter is the quantum mechanical theoretical prediction of the result of experimental observations, which are indeed real numbers! Read in reverse order, one can say that physical observables can only be represented by operators which are Hermitian (i.e. by operators that satisfy Eq. (2.23)) since these operators have real eigenvalues and the result of experimental measurements is a real quantity.

Commutators are very powerful operations and play a central role in quantum mechanics. The most useful properties of commutators are:

1. The commutator is a linear operation, i.e.:

$$\left[k\hat{A},\hat{B}\right]=k\left[\hat{A},\hat{B}\right] \tag{2.24}$$

with k a constant. And:

$$\left[\hat{A}+\hat{C},\hat{B}\right]=\left[\hat{A},\hat{B}\right]+\left[\hat{C},\hat{B}\right] \tag{2.25}$$

2. The commutator of a product of two operators is:

$$\left[\hat{A}\hat{C},\hat{B}\right]=\hat{A}\left[\hat{C},\hat{B}\right]+\left[\hat{A},\hat{B}\right]\hat{C} \tag{2.26}$$

3. The commutator is antisymmetric with respect to the exchange of the two operators, i.e.:

$$\left[\hat{A},\hat{B}\right]=-\left[\hat{B},\hat{A}\right] \tag{2.27}$$

4. The commutator of an operator with itself is always zero, i.e.:

$$\left[\hat{A},\hat{A}\right]=0 \tag{2.28}$$

The result in Eq. (2.21), which I report here in a neat line so to stress its importance:

TIME EVOLUTION OF THE EXPECTATION VALUE

$$\frac{d}{dt}\langle\hat{\mathrm{O}}\rangle=-\frac{i}{\hbar}\langle[\hat{\mathrm{O}},\hat{\mathrm{H}}]\rangle \tag{2.29}$$

is truly remarkable under various aspects, not least for its similitude with the most elegant rendering of classical mechanics due to Poisson and discussed later in this book.

If you have seen what was briefly discussed in this chapter for the first time in your studies, I imagine you feel quite overwhelmed by the depth of the new concepts introduced, the *oddity* of some of the conclusions, and the swift way of delivery that uses statements that may look like coming out of the blue. If you do feel so, or you have felt so when you were attending lectures on the topics, do not worry, it is quite normal, I felt the same and so did the vast majority of us who come across this material. In fact, I could now proceed along the same lines of most quantum mechanics books and courses, discussing in more depth the concept of operators, defining their properties, introducing the Dirac notations of *bras* and *kets*, talk about quantum states and their connection with the wavefunction, derive the predictions of quantum mechanics for a free particle, a particle in a box, a quantum spring, all the way to the description of an hydrogen atom and perhaps beyond. However, all of this requires the reader to be familiar with many concepts of classical physics. Neglecting to learn (or teach) those classical concepts will only make it more difficult to understand quantum mechanics. This is the very point of this book that I now begin to unravel by introducing, in the next few chapters, the concepts of classical mechanics in an effort to help the student to make sense, quicker and deeper, of the quantum mechanics discussed above and of what comes beyond. In reading the rest of the book, I want you to keep in mind Eq. (2.29) since, as I develop the classical mechanics from Newton to Poisson formulations, I aim to show you the classical counterpart of this very equation.

CHAPTER 3

Kinematics

IN THIS CHAPTER, I consider the concept of *motion* and the quantities used in *classical mechanics* to characterize motion. As briefed above, classical mechanics is a theory in physics that deals with the motion of *macroscopic* objects. It is typically divided into two branches: *kinematics* and *dynamics*. The first branch deals with the motion of macroscopic objects without considering the forces that cause such motion. The second deals with those forces. Note, once again, that the word *macroscopic* is crucial since, as discussed above, classical mechanics fails to fully describe the motion of *microscopic* objects.

Perhaps the most basic quantity to consider in mechanics is the object's *position*. It is as basic as important because the motion of an object is essentially a specification of its position at each instant in time, also known as the object's *trajectory*. Position is specified with respect to a reference frame. In a Cartesian reference frame, there are three axes that are orthogonal to each other and point along the directions known as x, y and z. In such a frame, one would indicate the position of an object with the position vector $\vec{r}(t)$:

POSITION VECTOR

$$\vec{r}(t)=\begin{pmatrix} x(t) \\ y(t) \\ z(t) \end{pmatrix} \tag{3.1}$$

DOI: 10.1201/9781003459781-3

The second most basic quantity in mechanics is the object's *velocity*, also a vector. Velocity is defined as the change of the object's position with time. In mathematics, the changes of a function with respect to a variable is given by the function's *derivative* with respect to that variable. Therefore, the velocity vector is expressed as the derivative of the position with respect to time, i.e.:

VELOCITY VECTOR

$$\vec{v}(t)=\frac{d}{dt}\vec{r}(t)=\begin{pmatrix}\frac{d}{dt}x(t)\\ \frac{d}{dt}y(t)\\ \frac{d}{dt}z(t)\end{pmatrix} \tag{3.2}$$

Through the book, I would like the reader to familiarize with some common notation in physics: for example, the time derivative of a function is often indicated by a *dot* over the function's symbol; a subscript is also commonly used to indicate a specific component of a vector as in:

$$\begin{aligned}v_x(t)&=\frac{d}{dt}x(t)=\dot{x}(t)\\ v_y(t)&=\frac{d}{dt}y(t)=\dot{y}(t)\\ v_z(t)&=\frac{d}{dt}z(t)=\dot{z}(t)\end{aligned} \tag{3.3}$$

A third basic quantity in mechanics is the object's *acceleration*. This quantity describes how velocity changes with time. As seen before, changes with time are expressed through derivatives hence, we write:

ACCELERATION VECTOR

$$\vec{a}(t)=\dot{\vec{v}}(t) \tag{3.4}$$

with components:

$$\begin{aligned} a_x(t) &= \dot{v}_x(t) \\ a_y(t) &= \dot{v}_y(t) \\ a_z(t) &= \dot{v}_z(t) \end{aligned} \tag{3.5}$$

By comparing Eq. (3.2) with Eq. (3.4), you see how the acceleration can also be thought as the second time derivative of position with respect to time, i.e.:

$$\vec{a}(t) = \ddot{\vec{r}}(t) \tag{3.6}$$

Note the use of *two dots* to indicate the second time derivative.

In mathematics, the sum of the squares of all components of a vector is called the *magnitude* of the vector, which is also the result of the inner product of a vector by itself. The square root of the magnitude is called the *norm* of the vector. The norms of the three vectors introduced above to represent position, velocity, and acceleration are:

$$\begin{aligned} |\vec{r}(t)| &= \sqrt{x^2(t) + y^2(t) + z^2(t)} \\ |\vec{v}(t)| &= \sqrt{v_x^2(t) + v_y^2(t) + v_z^2(t)} \\ |\vec{a}(t)| &= \sqrt{a_x^2(t) + a_y^2(t) + a_z^2(t)} \end{aligned} \tag{3.7}$$

The first quantity in Eq. (3.7) is also known as the *Euclidean distance* of the object from the origin of the reference frame used to report its position; the second quantity is known as *speed*; and the third is the *strength* of acceleration.

Depending on the exact form of velocity and acceleration, a motion assumes some peculiar characteristics. Motion is therefore classified through one or more adjectives that specify these characteristics. The main goal of kinematics is to determine the *equation of motion*, i.e., the equation describing the position of an object as a function of its velocity and acceleration.

I discuss below a few important types of motion and their equation of motion:

- *Uniform linear motion*: this is the motion of an object moving along a linear trajectory with constant velocity (i.e. null acceleration). The equation of motion can be derived by formulating the mathematical rendering of the statement above as:

$$\vec{a}(t) = \vec{0} \tag{3.8}$$

Assuming the motion happens along the x-axis only (to simplify the description) and using Eq. (3.5), the condition above can be written as:

$$\dot{v}_x(t) = 0 \tag{3.9}$$

Eq. (3.9) is a first-order differential equation whose solution is:

$$v_x(t) = \mathrm{v} \tag{3.10}$$

with v a constant quantity. Eq. (3.10) can be inserted into Eq. (3.2) to find:

$$\dot{r}_x(t) = \mathrm{v} \tag{3.11}$$

which is also a first-order differential equation whose solution gives:

$$r_x(t) = r_x(0) + \mathrm{v}t \tag{3.12}$$

Equation (3.12) is the equation of motion that describes a uniform linear motion. It is clearly the equation for a line and describes the position of the object as a function of a constant velocity and no acceleration. The motion can be visualized in a space–time diagram as shown in Figure 3.1.

- *Uniformly accelerated linear motion*: this is the motion of an object moving along a linear trajectory with a constant acceleration, here indicated by the constant $\vec{\mathrm{a}}$. The equation of motion can be derived by formulating the mathematical rendering of the statement above as:

$$\vec{a}(t) = \vec{\mathrm{a}} \tag{3.13}$$

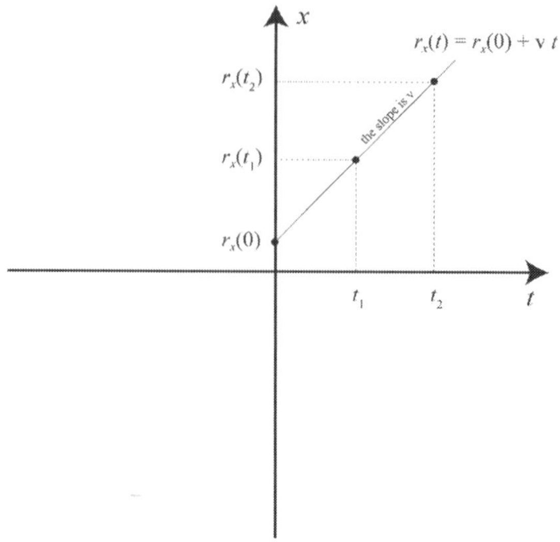

FIGURE 3.1 The space–time diagram of a uniform linear motion.

Assuming, again for the sake of simplicity, that the motion happens along the x-axis and using Eq. (3.5), I rewrite the condition above as:

$$\dot{v}_x(t) = \mathrm{a}_x \tag{3.14}$$

a first-order differential equation with solution:

$$v_x(t) = v_x(0) + \mathrm{a}_x t \tag{3.15}$$

Equation (3.15) can be inserted into Eq. (3.2) to produce a new differential equation for the position, namely:

$$\dot{r}_x(t) = v_x(0) + \mathrm{a}_x t \tag{3.16}$$

which is solved to give:

$$r_x(t) = r_x(0) + v_x(0)t + \frac{1}{2}\mathrm{a}_x t^2 \tag{3.17}$$

Equation (3.17) is the equation of motion describing an object in uniformly accelerated linear motion. The motion can be visualized in a space–time diagram as shown in Figure 3.2.

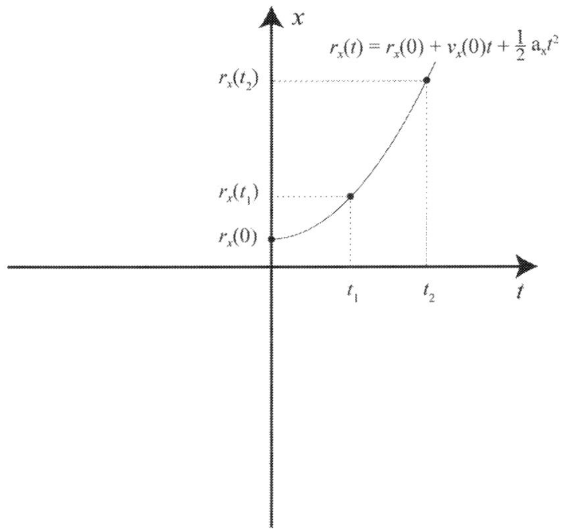

FIGURE 3.2 The space–time diagram of a uniformly accelerated linear motion.

- *Uniform circular motion*: this is the motion of an object moving on a circular trajectory of constant radius and with constant angular velocity (or zero angular acceleration). This motion is not too different from the uniform linear motion when the following facts are considered: i) a circular motion is bidimensional, this is because a circle is a bidimensional figure. The circle can be clearly tilted in a 3D space but, for simplicity, I assume it lies on the *xy* plane; ii) when an object is moving along a circular path of radius *R* (which I assume constant for now, and generalize it to the time-dependent radius case in Chapter 6), its instantaneous position becomes a function of the angle between one of the axes (any of the two) and the direction of the vector connecting the object with the frame origin. Hence, the position of a rotating object can be fully represented as the value that such angle assumes with time, i.e.:

ANGULAR COORDINATE

$$\theta(t) \tag{3.18}$$

The pair of coordinates, R and θ, are called *polar coordinates*. Note that the circular motion is the first example where we find it more convenient to abandon Cartesian coordinates for a different set (polar coordinates in this case). I will discuss about coordinates and change of coordinates in more detail in Chapter 6. The rate at which the angle $\theta(t)$ changes with time is the *angular velocity* that, in analogy with Eq. (3.2), can be expressed as:

ANGULAR VELOCITY

$$\omega(t) = \dot{\theta}(t) \tag{3.19}$$

Finally, the rate of change of the angular velocity is the *angular acceleration*, i.e.:

ANGULAR ACCELERATION

$$\alpha(t) = \dot{\omega}(t) = \ddot{\theta}(t) \tag{3.20}$$

Putting all together, when an object moves in a circular path with constant angular velocity, the equation of motion has the form:

$$\alpha(t) = \dot{\omega}(t) = 0 \tag{3.21}$$

with solution:

$$\omega(t) = \omega \tag{3.22}$$

where ω is the constant angular velocity. Equation (3.22) can be used in Eq. (3.19) to derive:

$$\begin{gathered} \dot{\theta}(t) = \omega \\ \theta(t) = \theta(0) + \omega t \end{gathered} \tag{3.23}$$

which is analogous to Eq. (3.12) but with the position vector replaced by the angular position coordinate. Incidentally, from Eq. (3.23), the total time needed for the object to describe an angle of 2π (meaning to go around the full circle once) is known as the period of the motion:

$$T = \frac{2\pi}{|\omega|} \tag{3.24}$$

Since, in a period, the object rotates describing a circumference of length $2\pi R$, one can derive the object linear speed as:

$$v = \frac{2\pi R\omega}{2\pi} = R\omega \tag{3.25}$$

The linear velocity is perpendicular to the radius of the circle at any instant of time. Moreover, the circle itself is on a plane which is perpendicular to the direction of the angular velocity pseudovector. Hence, the correct direction of the linear velocity vector is perpendicular to both the radius and the angular velocity; this fact can be expressed by means of the *vector cross product* as in:

LINEAR VELOCITY

$$\vec{v}(t) = \vec{\omega} \times \vec{R} \tag{3.26}$$

Note that the direction of the radius vector $\vec{R}$ points from the origin to the object.

- An *oscillating motion* or *oscillation* is the motion of an object that oscillates between a maximum and a minimum while traveling along a direction. This motion can be seen as the trajectory spanned by the projection along a given direction of an object in uniform circular motion. This is better illustrated in Figure 3.3.

  Having said so, the equation of motion for an oscillating object can be derived with the use of some simple trigonometry. The aim is to find the trajectory of the position vector $r_x(t)$ in Figure 3.3 resulting from the projection of the instantaneous position of the object as it rotates around the circle. Since the radius of the rotating motion is fixed to the value R, I can write:

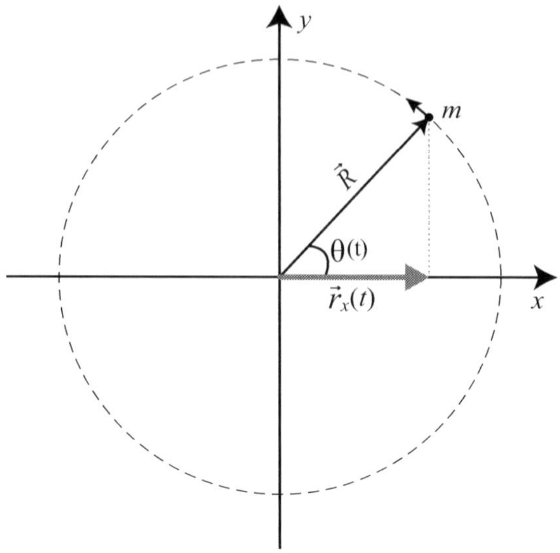

FIGURE 3.3 The projection onto the x-axis of the instantaneous position of an object in uniform circular motion describes an oscillating motion on either of the x- or y-axes.

$$r_x(t) = R\cos(\theta(t)) \tag{3.27}$$

But, for an object rotating in uniform circular motion, the angular position changes with time according to Eq. (3.23), hence:

$$r_x(t) = R\cos(\theta(0) + \omega t) \tag{3.28}$$

Equation (3.28) is the equation of motion for an oscillating object described as the projection along the x axis of the object rotating along a circular path of radius R and with constant angular velocity ω. The object's trajectory oscillates between $+R$ and $-R$ (this is known as the *oscillation amplitude*) at the angular frequency:

$$\nu = \frac{\omega}{2\pi} \tag{3.29}$$

The quantity $\theta(0)$ is known as the *oscillation phase*. Finally, the *oscillation period*, defining how long does it takes to complete one full oscillation, is given by (see Eq. (3.24)):

$$T = \frac{2\pi}{\omega} \tag{3.30}$$

As a final remark, note how the most basic players in the implementation of classical mechanics, i.e., position, velocity and acceleration are fundamentally different from those of quantum mechanics (see Chapter 2), where all such information is contained into the wavefunction and extracted by the action of an operator.

CHAPTER 4

Newtonian Mechanics

IN THE PREVIOUS CHAPTER, I have discussed the concept of motion showing how to derive the equation of motion with the use of a few simple definitions and differential equations. In this chapter, I shall address the questions: what puts objects in motion? And, if an object is in motion along a trajectory, how could its trajectory be changed? The short answer to both questions is: with the use of a *force*! Hence, the force is connected, in a way still to be discussed, to motion. If an object, say a football, rests on the floor and one kicks it with some force, it evidently will move. Would it move indefinitely? No, it will come to a stop, sooner or later depending on the strength of the force applied, and other factors. Does it stop spontaneously? One would be tempted to answer yes. Perhaps most, if not all, people born before Newton's work on mechanics would probably answer so. But contrarily to what it may look obvious to some, stopping a moving object does again require the use of a force. And, by the way, a force is also required to change the trajectory of a moving object.

Force and motion have been connected by Isaac Newton in what is known as Newton's second law:

NEWTON'S SECOND LAW

$$\vec{F}(t) = m\vec{a}(t) \tag{4.1}$$

where such connection is essentially expressed in terms of the acceleration discussed in Chapter 3. It will soon appear clear to you that it would make

DOI: 10.1201/9781003459781-4

more sense to label this law as Newton's first law because what is actually known as Newton's first law is a particular case of Eq. (4.1), but this is an unimportant formality.

Force, as well as position, velocity and acceleration encountered before, is a vector with three components. The proportionality constant between force and acceleration is the object's *mass*, denoted by m. Applying a force to an object of a given mass results in producing an acceleration. The heavier the object, the stronger will be the force required to achieve the desired acceleration. Using what we learnt in Chapter 3, we can rewrite Eq. (4.1) in terms of time derivatives as:

$$\vec{F}(t) = m\dot{\vec{v}}(t) \tag{4.2}$$

I shall now solve Eq. (4.2) for some cases you are already familiar with.

- An object moving with a certain velocity, but no force applied. For such a system, Eq. (4.2) becomes:

$$m\dot{\vec{v}}(t) = \vec{0} \tag{4.3}$$

A similar equation, a first-order differential equation, has been already encountered in the previous chapter and its solution is:

$$\vec{v}(t) = \vec{\mathrm{v}} \tag{4.4}$$

where $\vec{\mathrm{v}}$ is a constant velocity vector. Equation (4.4) is known as Newton's first law; it essentially states: *if an object is in motion with a certain velocity and there are no forces acting on it, it will continue its motion at constant velocity.* There you see why I have earlier written that Newton's first law is a particular case of his second law. Moreover, since the velocity is the time derivative of position, the trajectory of an object moving without a force applied is:

$$\begin{aligned} \dot{\vec{r}}(t) &= \vec{\mathrm{v}} \\ \vec{r}(t) &= \vec{r}(0) + \vec{\mathrm{v}}\,t \end{aligned} \tag{4.5}$$

which is the equation of motion for an object in uniform linear motion (see Chapter 3).

- In a second example, I introduce a constant force acting on the object. For the sake of simplicity, I assume the force is applied only along the x axis. Newton's second law for this system is spelled out as:

$$\begin{aligned} m\dot{v}_x(t) &= F_x \\ m\dot{v}_y(t) &= 0 \\ m\dot{v}_z(t) &= 0 \end{aligned} \tag{4.6}$$

Note that the components of the motion along the y and z direction fall within the case described by Eq. (4.5). The equation of motion along the x direction is obtained by solving the first-order differential equation in the first line of Eq. (4.6), which gives:

$$\begin{aligned} \dot{v}_x(t) &= \frac{F_x}{m} \\ v_x(t) &= v_x(0) + \frac{F_x}{m}t \end{aligned} \tag{4.7}$$

or, in terms of the object's position:

$$\begin{aligned} \dot{x}(t) &= v_x(0) + \frac{F_x}{m}t \\ x(t) &= x(0) + v_x(0)t + \frac{F_x}{2m}t^2 \end{aligned} \tag{4.8}$$

which is the equation of motion for an object in uniformly accelerated motion previously encountered in Chapter 3.

- Finally, I consider the case of a *restoring force*, such as, for example, the force acting in a spring, or a pendulum, or between atoms involved in atomic bonds, to cite a few real cases. The simplest form of a restoring force, again considering a single direction for the sake of simplicity, is the following:

$$F_x = -kx(t) \tag{4.9}$$

It describes a force whose strength is directly proportional to the position of the object along the x axis. In Eq. (4.9), k is a proportionality constant and regulates the strength of the force; the minus sign accounts for a force pointing against the direction of motion; in this way, the more the object is displaced along the x axis the stronger is the force that pulls the object back where it was — hence the name *restoring force*. In this case, Newton's second law gives:

$$\ddot{x}(t) = -\frac{k}{m}x(t) \tag{4.10}$$

which is a second-order differential equation. Think of it as asking for which mathematical function can describe the trajectory $x(t)$ when its second derivative is equal to the function itself (other than a minus sign and a multiplicative factor). Recalling some basic mathematics, the functions that remain the same after two consecutive derivatives are the trigonometric *sine* and *cosine* functions. It is therefore possible to verify that a generic solution of Eq. (4.10) is:

$$x(t) = \mathrm{a}\ cos\left(\sqrt{\frac{k}{m}}t\right) + \mathrm{b}\ sin\left(\sqrt{\frac{k}{m}}t\right) \tag{4.11}$$

with a and b two constants. The trajectory in Eq. (4.11) describes an oscillating motion with frequency:

$$\omega = \sqrt{\frac{k}{m}} \tag{4.12}$$

A final but very important note before concluding this lecture: forces in Nature can be classified as *conservative* and *non-conservative* (the meaning of this distinction will be clearer in the next chapter). Conservative forces, can be thought as generated by a *potential energy function* and are written as:

CONSERVATIVE FORCES

$$\vec{F}(\vec{r}) = -\frac{\partial V(\vec{r})}{\partial \vec{r}} \tag{4.13}$$

For example, the potential function generating the restoring force encountered above is:

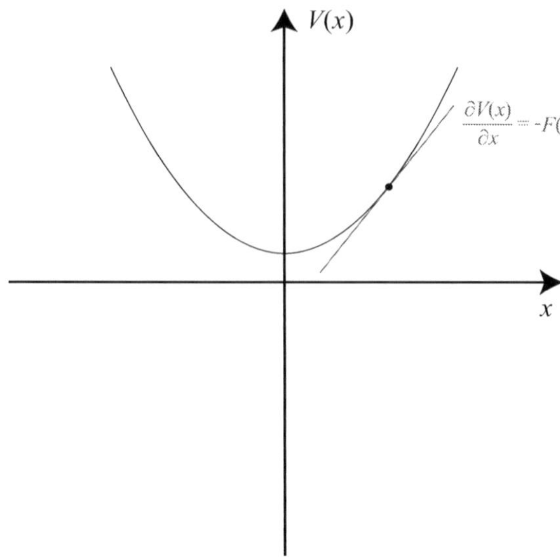

FIGURE 4.1 The force as the derivative of a potential function illustrated in the case of a harmonic potential. The derivative of the potential at any point is the line tangent to the curve in that point and represent minus the force acting on the point in that location.

$$V(x) = \frac{1}{2}kx^2 \tag{4.14}$$

which is the equation for a parabola. This potential function is generally called a *harmonic potential.* The relationship between force and potential in the case of conservative forces is illustrated in Figure 4.1:

This is clearly only one of the possible potential functions encountered in Nature. Figuring out the potential function giving rise to a certain conservative force is not always an easy task. Incidentally, if the force is known, then the potential function can be derived by rearranging and integrating both sides of Eq. (4.13) as in:

POTENTIAL ENERGY

$$V(\vec{r}) = -\int \vec{F}(\vec{r})d\vec{r} \tag{4.15}$$

An important remark on the meaning of Eq. (4.13): *a conservative force is generated by a potential energy function in the way that the force is always directed such as to push the object towards a lower potential energy status.*

CHAPTER 5

Energy

IT IS PROBABLY FAIR to start this chapter by asking: what is energy? A question that may give rise to plenty of philosophical discussion but with a rather precise scientific answer: *energy is the ability to do work.* Often the word *energy* is accompanied by an adjective that tries to better describe which form of energy is discussed. For instance, you may have heard about chemical energy, electrostatic energy, internal energy and several more. Ultimately, however, all those forms of energy can be fitted into one of these two fundamental categories: *potential energy* and *kinetic energy.*

Potential energy has been briefly encountered in Chapter 4 as the generator of conservative forces. Because forces make objects move and interact with each other, potential energy can be associated with the relative position of objects in space and the mutual forces between them. If two objects are in some position in space and can interact through a given force, say for example, two masses possessing an electric charge are at some distance from each other, then there is an electric force between them and such force is due to an electrostatic potential, i.e., there is some potential energy somehow stored in that system of two charged masses. Note that the potential energy is not a conserved quantity in the sense that once the force is exerted and the objects move, then the potential energy changes. And we know from Eq. (4.13) that the change is such that the potential energy diminishes.

Kinetic energy is the energy associated with motion. If an object moves with a velocity, v, then it has some kinetic energy, T, which is given by:

DOI: 10.1201/9781003459781-5

KINETIC ENERGY

$$\begin{aligned} T &= \frac{1}{2}m(\vec{v}\cdot\vec{v}) \\ &= \frac{1}{2}mv^2 \end{aligned} \tag{5.1}$$

where the *dot* symbol placed between two vectors returns their scalar product. The sum of potential and kinetic energy gives the *total energy* of the object, namely:

TOTAL ENERGY

$$E = T + V(\vec{r}) = \frac{1}{2}m(\vec{v}\cdot\vec{v}) - \int \vec{F}(\vec{r})\cdot d\vec{r} \tag{5.2}$$

The total energy of a system, whether it is made by a single object or many interacting ones, is conserved in *closed systems.* A system is said to be *closed* when it does not exchange neither mass nor energy with the rest of the universe. Saying that a quantity is conserved essentially means that such quantity does not vary with time. Therefore, to prove that the total energy of a closed system is conserved it requires calculating the time derivative of the total energy and verify that it is equal to zero. This is done by calculating:

$$\dot{E} = \dot{T} + \dot{V}(\vec{r}) \tag{5.3}$$

which I evaluate below in two separate steps. First, I calculate the derivative of the kinetic energy to obtain:

$$\begin{aligned} \dot{T} &= \frac{d}{dt}\left(\frac{1}{2}m(\vec{v}\cdot\vec{v})\right) \\ &= m(\vec{v}\cdot\dot{\vec{v}}) \\ &= m(\vec{v}\cdot\vec{a}) \end{aligned} \tag{5.4}$$

and then, I calculate the derivative of the potential energy to obtain:

$$\begin{aligned}\dot{V} &= \frac{d}{dt}V(\vec{r}) \\ &= \frac{\partial V(\vec{r})}{\partial \vec{r}} \cdot \frac{d\vec{r}}{dt} \\ &= -\vec{F}(\vec{r}) \cdot \vec{v}\end{aligned} \tag{5.5}$$

where, in the last equality, I used the results in Eqs. (4.13) and (3.2). Feeding back Eqs. (5.4) and (5.5) into Eq. (5.3), I obtain:

$$\dot{E} = \left(m\vec{a} - \vec{F}(\vec{r})\right) \cdot \vec{v} = 0 \tag{5.6}$$

where I have invoked Newton's second law (Eq. (4.1)) to balance out the two terms in the round parenthesis. The result in Eq. (5.6) demonstrates that the total energy is a conserved quantity in a closed system. When an object (or a system of objects) moves in a region with a given potential, the object's kinetic energy diminishes by some amount while its potential energy increases by the same amount, and *vice versa*.

It is quite the time to introduce another fundamental quantity in mechanics, the *linear momentum*. This is defined as:

LINEAR MOMENTUM

$$\vec{p} = m\vec{v} \tag{5.7}$$

The linear momentum is connected to the force as can be easily verified by recalling Eq. (4.1) and writing:

$$\begin{aligned}\vec{F} &= m\vec{a} \\ &= m\frac{d\vec{v}}{dt} \\ &= \frac{d(m\vec{v})}{dt} \\ &= \frac{d\vec{p}}{dt}\end{aligned} \tag{5.8}$$

Hence, *the rate of change of the linear momentum of an object is equal to the force acting on the object*. Moreover, the definition of the linear momentum is often used in the equation for the kinetic energy to obtain:

$$
\begin{aligned}
T &= \frac{(\vec{p} \cdot \vec{p})}{2m} \\
&= \frac{p^2}{2m}
\end{aligned}
\tag{5.9}
$$

which is equivalent to Eq. (5.1) and extensively used in mechanics.

Having introduced the concept of linear momentum, it becomes now possible to formulate Newton's third law as: *if two objects interact with each other, the force exerted by the first object on the second is equal in strength but opposite in sign to the force exerted by the second object on the first one.* In mathematics, this is written as:

NEWTON'S THIRD LAW

$$
\vec{F}_{12} = -\vec{F}_{21} \tag{5.10}
$$

Newton's third law is alternatively, and more often, expressed with the following sentence: *for every action in nature there is an equal and opposite reaction.* Interestingly, this law leads to another fundamental conservation law in physics, i.e., the *conservation of linear momentum.* To derive this conservation law, I proceed in the same way as I did for the conservation of energy, i.e., by showing that the time derivative of the linear momentum (a.k.a. its rate of change) is null in a closed system. Let me assume that object 1, with mass m_1 has linear momentum $\vec{p}_1 = m_1\vec{v}_1$ and, object 2 has mass m_2 and linear momentum $\vec{p}_2 = m_2\vec{v}_2$. If the two objects interact, Newton's third law can be written as:

$$
\begin{aligned}
\vec{F}_{12} &= -\vec{F}_{21} \\
\frac{d\vec{p}_1}{dt} &= -\frac{d\vec{p}_2}{dt} \\
\frac{d}{dt}(\vec{p}_1 + \vec{p}_2) &= 0
\end{aligned}
\tag{5.11}
$$

where, in the second line, I used the result in Eq. (5.8). Equation (5.11) demonstrates that the sum of the linear momenta of two interacting objects (you can extrapolate to many) is conserved, i.e., *the total linear momentum of a closed system does not change with time.* Furthermore, it can be shown that the linear momentum is always conserved if there are no external forces applied.

Linear momentum is not the only important form of momentum in classical mechanics. Another source of momentum is connected to the kinetic energy of a rotating body. To show this, I start by considering an object of mass, m, rotating about an axis (that does not change with time) at a distance $\vec{R}$ and angular velocity $\vec{\omega}$. Recalling the equation for the linear velocity of an object rotating on a circle (Eq. (3.26)):

$$\vec{v} = \vec{\omega} \times \vec{R} \tag{5.12}$$

and proceeding by inserting this latter result in the equation for the rotational kinetic energy (Eq. (5.1)), I obtain:

$$T_R = \frac{1}{2} m \left(\vec{\omega} \times \vec{R} \right)^2 = \frac{1}{2} m R^2 \omega^2 \tag{5.13}$$

By inspection of Eq. (5.13) and comparison with Eq. (5.1) some important similarities can be found. Firstly, the kinetic energy of a rotating object involves the square of its angular velocity while the kinetic energy of an object moving in a linear motion involves the square of the linear velocity. Secondly, the kinetic energy of a rotating object is proportional to the quantity mR^2 while the mass m appears in the kinetic energy of an object moving in a linear motion. The quantity mR^2 is known as the *moment of inertia (about the axis of rotation)* or *rotational inertia* or *angular mass:*

MOMENT OF INERTIA

$$I = mR^2 \tag{5.14}$$

The similitude between linear and circular motions does not stop here. Take, for example, the equation for the kinetic energy (Eq. (5.1)); by replacing the mass m with I and the linear velocity $\vec{v}$ with the angular velocity $\vec{\omega}$, one obtains:

$$T_R = \frac{1}{2} I \vec{\omega}^2 \tag{5.15}$$

which is the kinetic energy associated with a rotating object. This can alternatively be written in terms of an *angular momentum* $\vec{L}$ as in:

$$T_R = \frac{\vec{L}^2}{2I} \tag{5.16}$$

which is now the analogous of Eq. (5.9) where the linear momentum was used to write the kinetic energy of an object in linear motion. Note that, in the reasoning above I have implicitly defined the quantity angular momentum as:

ANGULAR MOMENTUM

$$\begin{aligned}\vec{L} &= I\vec{\omega} \\ &= mR^2\dot{\theta}\end{aligned} \tag{5.17}$$

The angular momentum is a pseudovector that has the same direction as the angular velocity, i.e., it points perpendicularly to the plane where the circular motion happens. Furthermore, by combining the results in Eqs. (5.12) and (5.14), the angular momentum can be expressed in terms of the linear momentum as:

$$\begin{aligned}\vec{L} &= m\left(\vec{R}\times\vec{\omega}\times\vec{R}\right) \\ &= m\left(\vec{R}\times\vec{v}\right) \\ &= \left(\vec{R}\times m\vec{v}\right) \\ &= \left(\vec{R}\times\vec{p}\right)\end{aligned} \tag{5.18}$$

where the *vector cross product* is used to return the correct direction of the angular momentum.

Angular momentum is also conserved, but only in the absence of a *torque*. For an object rotating describing a circular path, the torque is a vector that sits at the origin, and whose direction and magnitude results from the vector cross product between the radius $\vec{R}$ and the force applied to the object (see Figure 5.1), i.e.:

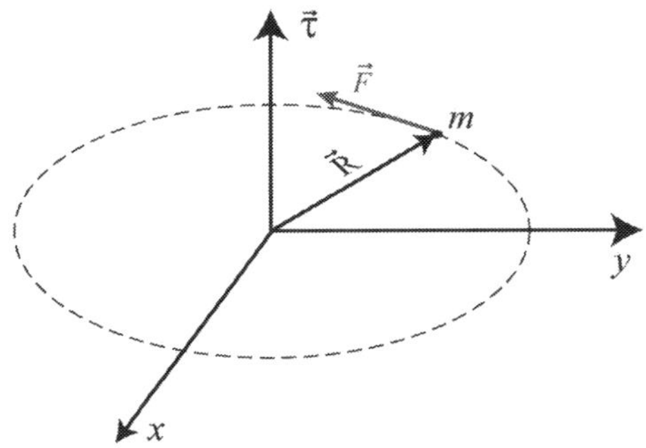

FIGURE 5.1 The torque arising when a force is applied to an object confined to move on a circular path.

TORQUE

$$\vec{\tau} = \left(\vec{R} \times \vec{F}\right) \tag{5.19}$$

Essentially, a torque represents the ability of a linear force to bring about a rotational motion. To demonstrate that the angular momentum of a rotating object is conserved in the absence of a torque, I proceed by taking the time derivative of the angular momentum calculated as:

$$\begin{aligned}\frac{d\vec{L}}{dt} &= \frac{d\left(\vec{R} \times \vec{p}\right)}{dt} \\ &= \frac{d\vec{R}}{dt} \times \vec{p} + \vec{R} \times \frac{d\vec{p}}{dt}\end{aligned} \tag{5.20}$$

The first term on the right-hand-side of Eq. (5.20) is zero because $\frac{d\vec{R}}{dt} = \vec{v}$ and $\vec{v} \times \vec{p} = 0$, since linear momentum and velocity are colinear. The second term on the right-hand-side of Eq. (5.20) contains the force (see Eq. (5.8)) hence I can conclude:

CONSERVATION OF ANGULAR MOMENTUM

$$\frac{d\vec{L}}{dt} = \vec{R} \times \vec{F} = \vec{\tau} \tag{5.21}$$

Therefore, if $\vec{\tau} = 0$ (i.e. there is no torque) then the angular momentum is conserved.

Note that for what is seen above, the kinetic energy is quadratic in the linear velocity for an object translating along a direction and quadratic in the angular velocity for an object rotating about an axis. In some cases, such as, for example, in an oscillating motion, the potential energy is also quadratic in the position (see Eq. (4.14)).

Bonus Section: The energy of a perfect gas (not essential to the aim of this book)

Since this chapter is about energy, I thought it may be nice to have a little diversion (from the aim of the book but not from classical physics) and introduce the energy in a perfect gas and the derivation of the notorious perfect gas state law.

A theorem known as the *equipartition theorem* (here not proved because this would require concepts of statistical mechanics) states: *each degree of freedom carries a contribution of* $\frac{1}{2}k_B T$ *to the averaged total energy of the system*, with $k_B = 1.38 \times 10^{-23}$ J K^{-1} being a constant known as the Boltzmann constant. In physics, a *degree of freedom*, is counted for each independent coordinate that describes the motion. If an object can translate in three dimensions and rotate about three axes, there will be three linear and three angular velocities terms in the total energy accounting for a total of six degrees of freedom. In addition to this, an object may have some internal degrees of freedom arising, for example, from the fact that parts of it can oscillate (harmonically) with respect to each other. Should this happen, we need to account for that many internal (or vibrational) degrees of freedom in addition to the ones arising from translational and rotational motions.

Consider a monoatomic gas made by a very large number of particles, N, and each atom of gas is considered point-like with mass m and no charge. Place such gas in a sealed container of volume V. A visualization of this is provided in Figure 5.2 to which the reader can refer for all steps below.

Because the gas is monoatomic, there could be no internal vibrational degrees of freedom, nor rotational ones because rotation has no meaning in a point-like representation. Hence, gas has only three translational degrees of freedom corresponding to velocities along the three directions of space. The total energy of one mole of such a monoatomic gas, as predicted by the equipartition theorem, is therefore:

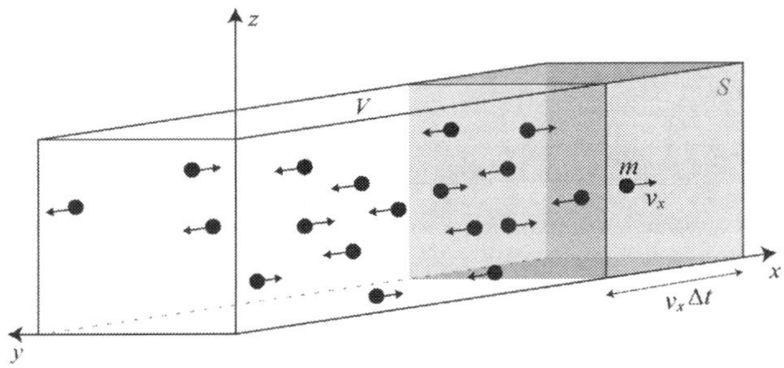

FIGURE 5.2 N atoms of a monoatomic gas made by identical particles of mass m trapped inside a box of volume V and yz surface S. Considering the sole x direction as an example, half the molecules move towards the surface S at the average velocity v_x and the other half move away from S at the average velocity $-v_x$

$$\langle E_{gas} \rangle = \frac{3}{2} N k_B T \ \ (\textit{monoatomic} \text{ gas}) \tag{5.22}$$

The gas has only kinetic energy and the total average kinetic energy of a gas made by identical atoms of mass m (and total mass $M = Nm$) is:

$$\langle E_{gas} \rangle = \frac{1}{2} \langle M v^2 \rangle \tag{5.23}$$

By combining the last two equations, we find out that the average atomic square velocity is proportional to the temperature of the gas (as you would have expected or heard of before):

$$\langle v^2 \rangle = \frac{3 k_B T}{m} \ \ (\textit{monoatomic} \text{ gas}) \tag{5.24}$$

The pressure of the gas is, by definition, the force exerted by the molecule of the gas on the walls of the container per unit of surface. This force is the result of molecular collisions and can be calculated using classical mechanics assuming the collisions are *elastic*. By saying that collisions are elastic one means that the energy is preserved during a collision. Since the energy of the gas is only of kinetic nature, then an *elastic collision* implies the conservation of the square velocity and, consequently, of the square linear momentum. Let me consider first the case in which the atoms are moving along the x direction to collide with a wall with surface S (see

Figure 5.2). From definition, the pressure exerted by the gas atoms on the surface S is:

$$P_x = \frac{F_x}{S} \tag{5.25}$$

Recalling Newton's second law, the force can be written as:

$$F_x = m\frac{\Delta v_x}{\Delta t} \tag{5.26}$$

where I have indicated the macroscopic changes in velocity and time with the symbol Δ. Since each atom hits the surface *elastically*, its incoming energy is preserved across the collision. Hence, the hitting atom bounces back with the same velocity, along the same axis, but travelling the opposite way than it was travelling before the collision with the wall surface. The change in its velocity upon the collision is therefore $\Delta v_x = 2v_x$ (it was moving at $+v_x$ towards the wall and, after colliding, it moves with velocity $-v_x$). The force exerted on the surface by any one single atom is therefore:

$$F_x = \frac{2mv_x}{\Delta t} \tag{5.27}$$

How many atoms of gas can collide with the wall in a time Δt? Assuming that on an average about half of the N atoms are travelling along $+x$ direction, then only the atoms occupying the volume $Sv_x\Delta t$ are close enough to the wall to hit it within a Δt (the shaded area in Figure 5.2). Therefore the number of atoms of gas colliding with the surface, N_c, is given by half the total number times the volume fraction of the atoms in the colliding volume over the total volume, i.e.:

$$N_c = \frac{N}{2}\frac{Sv_x\Delta t}{V} \tag{5.28}$$

The total force exerted on the surface by all the atoms moving along the positive x direction is:

$$\begin{aligned} F_x &= \frac{2mv_x}{\Delta t} \times \frac{N}{2}\frac{Sv_x\Delta t}{V} \\ &= \frac{mNSv_x^2}{V} \end{aligned} \tag{5.29}$$

Considering that velocity will be different for different atoms and therefore using its mean value, the pressure exerted is:

$$P_x = \frac{mN\langle v_x^2\rangle}{V} \tag{5.30}$$

In a more realistic tridimensional situation, velocities along the three directions must be considered. The average velocity along a particular direction has no reason to be different from the average velocity along a different one, thus:

$$\langle v_x^2\rangle = \langle v_y^2\rangle = \langle v_z^2\rangle = \frac{1}{3}\langle v^2\rangle \tag{5.31}$$

where $v^2 = v_x^2 + v_y^2 + v_z^2$. Hence, I can write:

$$P = \frac{1}{3}\frac{mN\langle v_x^2\rangle}{V} \tag{5.32}$$

The result in Eq. (5.32) can be combined with the one in Eq. (5.24) to obtain:

$$\begin{aligned} P &= \frac{1}{3}\frac{mN}{V}\frac{3k_BT}{m} \\ &= \frac{Nk_BT}{V} \end{aligned} \tag{5.33}$$

This is the very famous equation of state for an ideal gas, which you have probably encountered before. Let me make it more familiar to you by noting that the total number of molecules can be expressed as the number of moles, n, times the Avogadro's number, i.e., $N = nN_A$ with $N_A = 6.022\times10^{23}$ molecules per mole, and that the gas constant R is defined as:

$$R = N_A k_B = 8.31\ \mathrm{J\ K^{-1}\ mol^{-1}} \tag{5.34}$$

Combine these two expressions with Eq. (5.33) to obtain the best known:

STATE EQUATION FOR A PERFECT GAS

$$PV = nRT \tag{5.35}$$

Both sides of this equation have the dimension of Joules and represent the energy of the ideal gas. The adjective *ideal* is used to summarize all sort of approximations we had to make to derive this equation, and, in particular, the fact that the collisions can be considered elastic as well as the fact that we have not considered the possibility of interatomic interactions. Such ideality can strictly be assumed only for monoatomic gases at very low pressure, although the equation seems to hold quite well and is widely used also at room pressure.

CHAPTER 6

Lagrangian Mechanics

WHAT DISCUSSED SO FAR is essentially Newtonian mechanics whose formulation requires knowledge of the object's initial position and velocity, a total of six numbers (per object) since both position and velocity are three-dimensional vectors. Although I did not discuss this explicitly, if one has a system of N interacting (macroscopic) objects, then the dynamics of the systems can be fully predicted through Newtonian laws once the 6N initial positions and velocities are known. In the 18th century, though, the entire Newtonian mechanics was reformulated in a different, and often more convenient way. This formalism is known as *Lagrangian mechanics* and takes the name from the scientist Giuseppe Lodovico Lagrangia. Describing Lagrangian mechanics requires a step up in complexity and therefore it is better to prepare the ground in a more formal and structured way.

Firstly, I need to introduce the concepts of *configuration space* and *phase space*. As explained above, in Newtonian mechanics we use the set of positions and velocities to describe a physical system (may this be an object or a collection of interacting objects). The collection of all position vectors (one for each object) is known as *configuration space*, sometimes also known as the *position space*. The collection of all position vectors plus all momentum vectors is known as *phase space*. Configuration and phase spaces are illustrated in Figure 6.1 for a bidimensional case.

I shall now define the concept of *action* as a numerical quantity that describes how a system has evolved over time. Mathematically, the action is a *functional*, i.e., a function of functions. The mathematics of functionals

DOI: 10.1201/9781003459781-6

is the subject of the *calculus of variations* which, among others, establishes the mathematical tools to calculate maxima and minima of functionals. A crucial principle of this area of mathematics is the so-called *stationary-action principle* (a.k.a. *principle of least action*, a name that many physicists do not particularly like since it could be misleading). This principle states that: *trajectories are stationary points of the action functional for a system.* I understand it all sounds quite hard to understand right now, but I shall do my best to make it clearer as I discuss it further below. For now, let me clarify what a stationary point is. When dealing with a function, stationary points are those points in which the function does neither increase nor decrease. You may recall that these points are calculated by setting the function derivative to zero. In analogy, the stationary points of a functional are found by setting the functional derivative to zero. To make the point so far: we are interested in the trajectory of an object; the trajectory is proved to be the stationary point of an action functional; the stationary points of a functional are found by setting the functional derivative to zero. It only remains to say what the actual form of the action is and how to calculate its derivative!

The action is written as an integral over time between the start and final point of a trajectory. What goes inside the integral is the so-called *Lagrangian*, $\mathcal{L}$, which is given as the difference between the kinetic and the potential energy of the moving object:

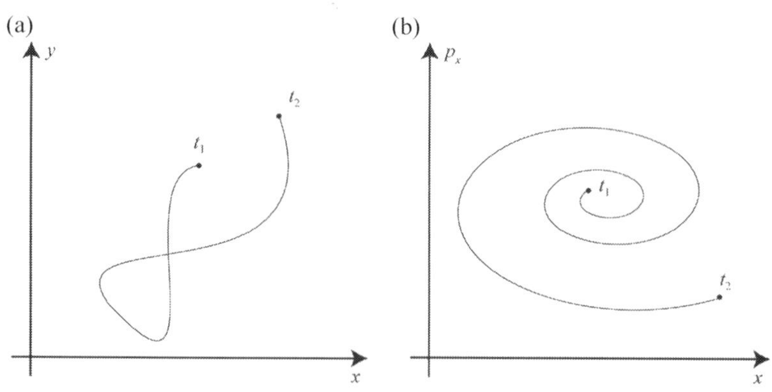

FIGURE 6.1 *a) an example trajectory of an object in a bidimensional xy configuration space; b) an example trajectory of an object in a phase space involving its position and momentum along the x direction.*

THE LAGRANGIAN

$$\mathcal{L}\left(\vec{r}(t),\dot{\vec{r}}(t),t\right)=T\left(\dot{\vec{r}}(t)\right)-V\left(\vec{r}(t),t\right) \tag{6.1}$$

The Lagrangian is a function of positions, $r(t)$, velocities, $\dot{r}(t)$, and time, t. Note, though, that the Lagrangian is always going to be a function of time through the time dependence of position and velocity, but it is important to figure out if the Lagrangian depends explicitly on time (read: whether the potential energy is time-dependent) since this fact has important repercussions on energy conservation, as will be discussed later. Given the definition above, the *action for a moving object* can be formalized as:

ACTION

$$\mathcal{A}\left(\vec{r}(t),\dot{\vec{r}}(t)\right)=\int_{t_1}^{t_2}\mathcal{L}\left(\vec{r}(t),\dot{\vec{r}}(t),t\right)dt \tag{6.2}$$

The stationary points of the action are found by setting to zero all derivatives with respect to all functions the action depends on (remember that the action is a functional and therefore it depends on functions, in this case positions and velocities that are functions of time). This condition is often written as:

$$\delta\mathcal{A}\left(\vec{r}(t),\dot{\vec{r}}(t)\right)=0 \tag{6.3}$$

I have read from several notable physicists that this is perhaps the closest equation we have to a theory of everything. Physicists have already written down the action whose stationary points reveal the equations for electrodynamics (i.e. Maxwell's equations), general relativity (i.e. Einstein's field equations), and quantum theory (Feynman's path integral).

To find the stationary points of an action may seem (and probably is) quite a difficult task. However, Euler and Lagrange have derived a very elegant and compact equation that ensure the stationarity of an action. This is known as the *Euler–Lagrange* equation and is written as:

EULER–LAGRANGE EQUATION

$$\frac{d}{dt}\frac{\partial \mathcal{L}}{\partial \dot{\vec{r}}} = \frac{\partial \mathcal{L}}{\partial \vec{r}} \tag{6.4}$$

where I have now omitted the time dependence of position and velocity to simplify the notation. What a remarkable equation, don't you agree? And, before providing examples to show the power of this equation, and widely, of Lagrange's formulation of mechanics, I want to reflect a bit on what the equations above really means. Imagine an object that occupies a given point in space at time t_1 and then moves to occupy another place at a later time t_2. Mechanics is all about predicting the trajectory it takes to move between these two points in time and space. Clearly, there are an infinite number of trajectories that can be taken, but only one is ultimately taken by the object. Lagrangian mechanics tells that the one taken must coincide with the trajectory for which the action is stationary. I tried to illustrate this graphically in Figure 6.2.

Imagine now that as the object moves along a given trajectory it experiences a given set of values of the Lagrangian (this is because the Lagrangian depends on — and therefore changes with — position, velocity, and time). The actual set of values experienced by the object is clearly different for different trajectories. Imagine then plotting the value of the Lagrangian experienced

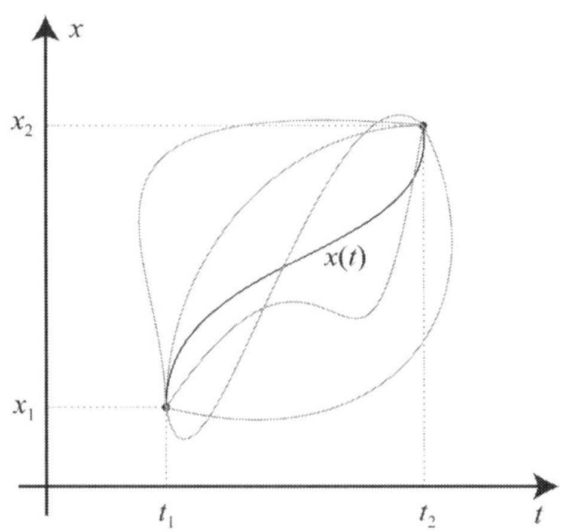

FIGURE 6.2 *To be found in x_2 at time $t_2$an object positioned in x_1 at time t_1 will only take the trajectory for which the "action" is stationary out of all the many possible trajectories.*

by the object while it moves along any of all possible trajectories. The action in Eq. (6.2) is the area under the curve in such plot. Finding stationary points is equivalent to minimizing this area. But, minimizing with respect to what variable? It is with respect to trajectories because the action is a function of trajectory functions. Hence, the object will move along the trajectory for which the action is stationary. An infinitesimally small change along this trajectory will not change the action, at a first-order approximation for the changed parameter. This is the same as saying that when we are at a stationary point of a function of a variable, an infinitesimally small change in that variable does not change the slope of the curve. Note, finally, that the action contains an integral into the future. We do not know yet where the object will be at time t_2 but it seems that the object knows ahead of time which trajectory to choose among the infinite possible choices! This way of setting the problem is more general than the approach taken by Newton and this generality will then reflect into a wider applicability and an easier approach of Lagrangian mechanics in hard mechanical problems.

I shall now demonstrate how the Euler–Lagrange equation is applied to derive the equation of motion and I am going to do this by revisiting some of the cases we have already explored in previous chapters using the Newtonian approach.

- I start again from the simplest case of an object in motion, no force applied. I have discussed this case in Chapter 4 using Newtonian mechanics to obtain the equation of motion reported in Eq. (4.4). Using the Lagrangian approach, I start by writing down the Lagrangian as:

$$\mathcal{L} = \frac{1}{2} m\left(\dot{\vec{r}} \cdot \dot{\vec{r}}\right) = \frac{1}{2} m\dot{r}^2 \tag{6.5}$$

  where there is only a kinetic energy term since no force means no potential energy! I proceed by calculating separately all terms required in Eq. (6.4):

$$\begin{aligned} \frac{\partial \mathcal{L}}{\partial \vec{r}} &= \vec{0} \\ \frac{\partial \mathcal{L}}{\partial \dot{\vec{r}}} &= m\dot{\vec{r}} \\ \frac{d}{dt}\frac{\partial \mathcal{L}}{\partial \dot{\vec{r}}} &= m\ddot{\vec{r}} \end{aligned} \tag{6.6}$$

And then put them all together to obtain:

$$m\ddot{\vec{r}} = \vec{0} \tag{6.7}$$

which is a second-order differential equation whose solution gives:

$$\dot{\vec{r}} = \vec{v} \tag{6.8}$$

Equation (6.8) clearly coincides with Newton's first law, previously encountered in Eq. (4.4).

- In a second example, I introduce a potential energy term with no explicit dependence on time, other than the one due to the position vector. The Lagrangian for this object is written as:

$$\mathcal{L} = \frac{1}{2}m\left(\dot{\vec{r}}\cdot\dot{\vec{r}}\right) - V\left(\vec{r}\right) \tag{6.9}$$

The three terms required in Eq. (6.4) are therefore:

$$\begin{aligned} \frac{\partial\mathcal{L}}{\partial\vec{r}} &= -\frac{dV\left(\vec{r}\right)}{d\vec{r}} = \vec{F}\left(\vec{r}\right) \\ \frac{\partial\mathcal{L}}{\partial\dot{\vec{r}}} &= m\dot{\vec{r}} \\ \frac{d}{dt}\frac{\partial\mathcal{L}}{\partial\dot{\vec{r}}} &= m\ddot{\vec{r}} \end{aligned} \tag{6.10}$$

where, in the first line, I used the definition of force given in Eq. (4.13). These three terms can be inserted into Eq. (6.4) to obtain:

$$\vec{F}\left(\vec{r}\right) = m\ddot{\vec{r}} \tag{6.11}$$

Equation (6.11) clearly coincides with Newton's second law, previously encountered in Eq. (4.1).

- In a third and final example, I discuss the case in which the potential is harmonic (earlier introduced in Eqs. (4.9) and (4.14)) so that the Lagrangian for the object becomes:

$$\mathcal{L}=\frac{1}{2}m\left(\dot{\vec{r}}\cdot\dot{\vec{r}}\right)-\frac{1}{2}k\left(\vec{r}\cdot\vec{r}\right) \tag{6.12}$$

The three terms required by the Euler–Lagrange equation are derived as:

$$\begin{aligned}\frac{\partial\mathcal{L}}{\partial\vec{r}}&=-k\vec{r}\\ \frac{\partial\mathcal{L}}{\partial\dot{\vec{r}}}&=m\dot{\vec{r}}\\ \frac{d}{dt}\frac{\partial\mathcal{L}}{\partial\dot{\vec{r}}}&=m\ddot{\vec{r}}\end{aligned} \tag{6.13}$$

and, once inserted into Eq. (6.4), led to:

$$m\ddot{\vec{r}}=-k\vec{r} \tag{6.14}$$

which is identical to what is found in Eq. (4.10) with the use of Newtonian mechanics.

I could continue with more examples, the most classical one would be the double pendulum since it demonstrates very clearly the enormous advantages of the Lagrangian approach over the Newtonian one. Rather, I prefer to address the interested reader to look out for such example in academic physics books or, better, to try it out approaching the problem mimicking the way I approached the cases above.

All the mechanics discussed so far was written using Cartesian coordinates, namely $\vec{r}(t)=\{x(t),y(t),z(t)\}$. However, there is nothing special with the use of those coordinates; they may be convenient to formalize and solve some problems, and less convenient to find solutions in others. It is fruitful, though, to familiarize with the use of *generalized coordinates*, typically indicated q_i where the subscript i indicates, at the same time, all directions of space and all objects in case the mechanical problem deals with systems made by a collection of objects. For example, if we are dealing with a single object, q_i represents the Cartesian set $\{q_x=x(t),q_y=y(t),q_z=z(t)\}$. If there are two of those objects, then q_i represents the set $\{q_{1x}=x_1(t),$ $q_{1y}=y_1(t),q_{1z}=z_1(t),q_{2x}=x_2(t),q_{2y}=y_2(t),q_{2z}=z_2(t)\}$, where $x_1(t)$ is the x coordinate of object 1, $x_2(t)$ is the x coordinate of object 2, and so on.

To be useful in solving problems with Lagrangian mechanics, generalized coordinates must satisfy three criteria:

1. be *independent*
2. form a *complete* set
3. be as many as the degrees of freedom

Generalized coordinates are *independent* if when all coordinates but one are fixed, then there is still free movement about the free coordinate. They are *complete* if the specified set of coordinates can locate all parts of the system, at all times. Finally, Lagrangian mechanics only applies to *holonomic* systems, i.e., systems where the generalized coordinates required to describe the system are as many as the system's degrees of freedom. The Euler–Lagrange equation written in generalized coordinates becomes:

EULER–LAGRANGE IN GENERALIZED COORDINATES

$$\frac{d}{dt}\frac{\partial\mathcal{L}\left(q_i(t),\dot{q}_i(t),t\right)}{\partial\dot{q}_i(t)}=\frac{\partial\mathcal{L}\left(q_i(t),\dot{q}_i(t),t\right)}{\partial q_i(t)} \tag{6.15}$$

For the sake of clarity, in the last equation I have explicitly written all variables and their time dependence. However, I will now drop, once again, the explicit time dependence to simplify the notation. Moreover, it is very important to remember that *there is a Euler–Lagrange equation for each generalized coordinate.*

At this point you may be asking why bothering to introduce generalized coordinates? I shall answer the question with an example. Consider an object that rotates on a plane following a circular trajectory of radius $R(t)$; to keep it very general the radius may be fixed or variable. As the object rotates, the angle $\theta(t)$ varies as shown in Figure 6.3. Consider now the following set of generalized coordinates $\{q_1 = R(t), q_2 = \theta(t)\}$. Firstly, do these two generalized coordinates satisfy the three criteria above? q_1 and q_2 are independent from each other: while fixing the radius, the object can still rotate in a circle of fixed radius; while fixing the angle, the object can still move in a line along the radius direction. q_1 and q_2 can obviously locate the object at any time. Finally, being constrained on a plane, the system

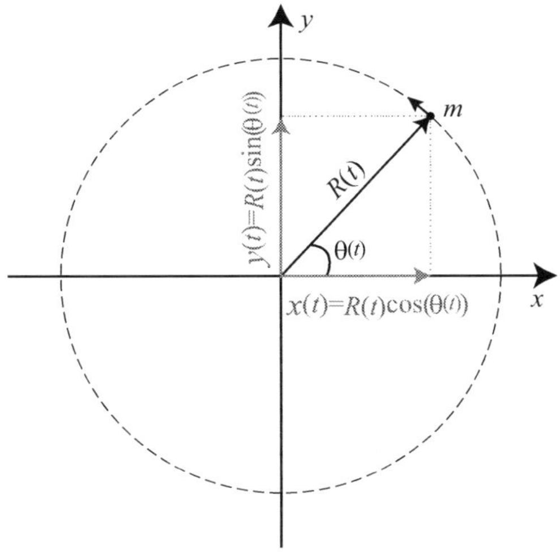

FIGURE 6.3 Trigonometric relationship between Cartesian and generalized coordinates for an object in uniform circular motion.

has two degrees of freedom and therefore two generalized coordinates are needed. Since all three criteria are satisfied, Lagrangian mechanics can be used to determine the equation of motion for the object.

To do so, the first task is to write the object's Lagrangian in terms of the two generalized coordinates chosen. This requires a bit of trigonometry, but the calculations are relatively easy. There is no potential energy, and the kinetic energy can be initially written down in terms of the more familiar Cartesian coordinates as in:

$$\begin{aligned} T &= \frac{1}{2} m \dot{\vec{r}}(t)^2 \\ &= \frac{1}{2} m\left(\dot{x}(t)^2 + \dot{y}(t)^2\right) \end{aligned} \tag{6.16}$$

With the use of trigonometry, the Cartesian coordinates occurring in Eq. (6.16) can be expressed in terms of the chosen generalized coordinates (see Figure 6.3) as:

$$\begin{aligned} x(t) &= R(t)cos(\theta(t)) \\ y(t) &= R(t)sin(\theta(t)) \end{aligned} \tag{6.17}$$

The time derivatives of these coordinates are calculated by applying the product rule. This will result in:

$$\begin{aligned}\dot{x}(t) &= \dot{R}(t)cos(\theta(t)) - R(t)sin(\theta(t))\dot{\theta}(t)\\ \dot{y}(t) &= \dot{R}(t)sin(\theta(t)) + R(t)cos(\theta(t))\dot{\theta}(t)\end{aligned} \tag{6.18}$$

The sum of their squares, which occurs in Eq. (6.16), is:

$$\dot{x}(t)^2 + \dot{y}(t)^2 = \dot{R}(t)^2 + R(t)\dot{\theta}(t)^2 \tag{6.19}$$

and the Lagrangian, now expressed in generalized coordinates, becomes:

$$\mathcal{L} = \frac{1}{2}m\dot{R}^2 + \frac{1}{2}mR^2\dot{\theta}^2 \tag{6.20}$$

where the time dependence of the coordinates has been dropped in order to simplify the notation. Since there are two generalized coordinates, there are two Euler–Lagrange equations, one for each coordinate, namely:

$$\frac{d}{dt}\frac{\partial \mathcal{L}}{\partial \dot{R}} = \frac{\partial \mathcal{L}}{\partial R} \tag{6.21}$$

and

$$\frac{d}{dt}\frac{\partial \mathcal{L}}{\partial \dot{\theta}} = \frac{\partial \mathcal{L}}{\partial \theta} \tag{6.22}$$

To solve Eq. (6.21), the following three terms are required:

$$\begin{aligned}\frac{\partial \mathcal{L}}{\partial R} &= mR\dot{\theta}^2\\ \frac{\partial \mathcal{L}}{\partial \dot{R}} &= m\dot{R}\\ \frac{d}{dt}\frac{\partial \mathcal{L}}{\partial \dot{R}} &= m\ddot{R}\end{aligned} \tag{6.23}$$

These can be inserted into Eq. (6.21) to obtain:

$$\ddot{R} = R\dot{\theta}^2 \tag{6.24}$$

Similarly, Eq. (6.22) requires the following three terms:

$$\begin{aligned}\frac{\partial \mathcal{L}}{\partial \theta} &= 0 \\ \frac{\partial \mathcal{L}}{\partial \dot{\theta}} &= mR^2\dot{\theta} \\ \frac{d}{dt}\frac{\partial \mathcal{L}}{\partial \dot{\theta}} &= mR^2\ddot{\theta}\end{aligned} \tag{6.25}$$

These can be inserted into Eq. (6.22) to obtain:

$$mR^2\ddot{\theta} = 0 \tag{6.26}$$

The problem under investigation can be seen as a generalized case of the uniform circular motion already treated in Chapter 3, this time with the motion happening on a circular path whose radius can change with time. It is easy to verify that when Eq. (6.26) is intended for a constant radius, it reduces to $\ddot{\theta} = 0$ which coincides with Eq. (3.21), earlier written for a uniform circular motion with fixed radius. Also note that in such a case, the R coordinate would not be part of the set of generalized coordinates since for a circular path of fixed radius the system has only one degree of freedom and therefore only one generalized coordinate is required.

Two final remarks will help, I hope, highlighting the power of formulating mechanics in generalized coordinates. Have a second look at the central line of Eq. (6.23), which I report here for convenience:

LINEAR MOMENTUM IN A LAGRANGIAN FORMALISM

$$\frac{\partial \mathcal{L}}{\partial \dot{R}} = m\dot{R} \tag{6.27}$$

Does it look familiar to you? Well, it is the radial component of the linear momentum since the time derivative of R is a radial velocity (see the definition of linear momentum given in Eq. (5.7)). And here is the central line of Eq. (6.25):

ANGULAR MOMENTUM IN A LAGRANGIAN FORMALISM

$$\frac{\partial \mathcal{L}}{\partial \dot{\theta}} = mR^2\dot{\theta} \tag{6.28}$$

Does it look familiar to you? Indeed, it is the angular momentum firstly introduced in Eq. (5.17). More on this equivalence and its importance will be discussed in the following chapter.

CHAPTER 7

Hamiltonian Mechanics

Looking at Eqs. (6.27) and (6.28) discussed in Chapter 6, a clear pattern can be recognized: both the linear and angular momentum result from the derivative of the Lagrangian with respect to velocity (linear and angular, respectively). In fact, in physics the derivative of the Lagrangian with respect to the velocity ($\dot{q}_i$) conjugated to a given generalized coordinate (q_i) is a very important quantity known as the *canonical momentum conjugated to the coordinate*:

CANONICAL MOMENTUM

$$p_i = \frac{\partial \mathcal{L}}{\partial \dot{q}_i} \tag{7.1}$$

Note that, given the definition above, the canonical momentum may or may not coincide with the linear or angular momentum discussed within Newtonian mechanics since the form of the canonical momentum is correlated with the actual form of the Lagrangian itself. The case I know best, because it is relevant to my field of expertise (nuclear spin dynamics), is the canonical momentum of an electron in a magnetic field where the Lagrangian is such that the canonical momentum contains the linear momentum plus another term proportional to the electron change and the magnetic vector potential.

Canonical momenta, as well as positions, are major players in a new representation of mechanics, the one introduced in the 19th century by

DOI: 10.1201/9781003459781-7

Willian R. Hamilton and that takes the name of *Hamiltonian mechanics*. In this representation, a new fundamental quantity is introduced and indicated by H, standing for *Hamiltonian*. H is defined as:

HAMILTONIAN

$$H = \sum_i p_i \dot{q}_i - \mathcal{L} \tag{7.2}$$

As seen in Chapter 6, the Lagrangian is a function of positions, velocities, and time. Once again, time dependence is secured through the time dependence of position and velocity but, often, an explicit time dependence of the potential energy part is also present. If we assume, for the time being, that the potential is time-independent, the total change in the Hamiltonian, indicated below as δH, would only depend on the infinitesimal changes in both position and velocity and therefore can be written as:

$$\delta H = \sum_i \dot{q}_i \delta p_i + p_i \delta \dot{q}_i - \frac{\partial \mathcal{L}}{\partial q_i} \delta q_i - \frac{\partial \mathcal{L}}{\partial \dot{q}_i} \delta \dot{q}_i \tag{7.3}$$

where I have used the product rule to take the derivative of the first term in Eq. (7.2). Using the definition of canonical momentum given in Eq. (7.1), the total change in the Hamiltonian simplifies to:

$$\delta H = \sum_i \dot{q}_i \delta p_i - \frac{\partial \mathcal{L}}{\partial q_i} \delta q_i \tag{7.4}$$

since the second and fourth terms in Eq. (7.3) sum up to zero due to Eq. (7.1).

Generally, δH can be directly calculated using the rules of derivation as in:

$$\delta H = \sum_i \frac{\partial H}{\partial p_i} \delta p_i + \frac{\partial H}{\partial q_i} \delta q_i \tag{7.5}$$

By comparing Eqs. (7.4) and (7.5), the following equations can be easily derived:

$$\begin{aligned}\frac{\partial H}{\partial p_i} &= \dot{q}_i \\ \frac{\partial H}{\partial q_i} &= -\frac{\partial \mathcal{L}}{\partial q_i}\end{aligned} \tag{7.6}$$

Moreover, using the definition of canonical momentum in Eq. (7.1), one can rewrite the Euler–Lagrange equation (see Eq. (6.15)) as:

$$\frac{d}{dt}p_i = \frac{\partial \mathcal{L}}{\partial q_i} \tag{7.7}$$

By putting together Eqs. (7.6) and (7.7), Hamilton formulated the following two equations:

HAMILTONIAN MECHANICS EQUATIONS

$$\begin{aligned}\frac{\partial H}{\partial p_i} &= \dot{q}_i \\ \frac{\partial H}{\partial q_i} &= -\dot{p}_i\end{aligned} \tag{7.8}$$

which render the whole classical mechanics of Newton and Lagrange in a very elegant way, and with such a remarkable symmetry!

As usual, I will use an example to illustrate the use and power of Hamiltonian mechanics. Consider, once again, the Lagrangian for an object with kinetic and potential energy, the one discussed in Eq. (6.9) but now rewritten in terms of the generalized coordinate q:

$$\mathcal{L} = \frac{1}{2}m\dot{q}^2 - V(q) \tag{7.9}$$

Recalling the form of the linear momentum, the last equation can be equivalently written as:

$$\mathcal{L} = \frac{p^2}{2m} - V(q) \tag{7.10}$$

Thus, the Hamiltonian for the system is readily found by applying Eq. (7.2) to obtain:

$$H = \sum_i p_i \dot{q}_i - \mathcal{L} = p\dot{q} - \frac{p^2}{2m} + V(q) \tag{7.11}$$

and because, from the very definition of linear momentum, we have:

$$\dot{q}_i = \frac{p}{m} \tag{7.12}$$

the Hamiltonian can be simplified as in:

$$H = \frac{p^2}{2m} + V(q) \tag{7.13}$$

Note how Eq. (7.13) coincides with the total energy of the object, as you can see by comparing it to Eq. (5.2). This is a fundamental feature of the Hamiltonian representation of mechanics: *the Hamiltonian returns the total energy of the system.*

Once the Hamiltonian is known for the system, one can use Hamilton's equations (Eq. (7.8)) to obtain the equation of motions. In the current example, the derivatives of the Hamiltonian in Eq. (7.13), taken with respect to position and momentum, are:

$$\begin{gathered} \frac{\partial H}{\partial p} = \frac{p}{m} \\ \frac{\partial H}{\partial q} = \frac{\partial V(q)}{\partial q} \end{gathered} \tag{7.14}$$

These can be inserted into Eq. (7.8) to obtain:

$$\begin{gathered} \frac{p}{m} = \dot{q} \\ \frac{\partial V(q)}{\partial q} = -\dot{p} \end{gathered} \tag{7.15}$$

The first line in Eq. (7.15) is the very definition of linear momentum, already encountered within Newtonian mechanics. The second line contains the derivative of the potential, which is the force $F(q)$. Recalling Eq. (4.13), the second equation in Eq. (7.15) can be rewritten as:

$$\begin{aligned} F(q) &= \dot{p} \\ &= \frac{d}{dt}(mv) \\ &= ma \end{aligned} \tag{7.16}$$

which is Newton's second law! It is worth noting that while Lagrangian mechanics would lead to the same result but through a second-order differential equation in the position, Hamiltonian mechanics yields two first-order differential equations, one for the position and one for the momentum. Hamiltonian mechanics is therefore better understood in the phase space, which I very briefly introduced in the opening of Chapter 6.

In Chapter 5, I discussed about the conservation law of the total energy. I shall now generalize such concept to find out the conditions that need to be satisfied for the total energy to be conserved. Conservation laws can be easily formalized within the framework provided by Hamiltonian mechanics. This is because the Hamiltonian is directly linked to the total energy and therefore the energy is conserved, i.e., it does not change with time, if the time derivative of the Hamiltonian is null. Formally, such derivative is written as:

$$\frac{d}{dt}H = \frac{d}{dt}\left(\sum_i p_i \dot{q}_i - \mathcal{L}(q_i, \dot{q}_i, t)\right) \tag{7.17}$$

where the explicit time dependence of the Lagrangian has now been conveniently reintroduced to facilitate the discussion. Since, as remarked many times before, position, velocity, and momentum are time-dependent quantities, the first term on the right-hand-side of Eq. (7.17) is:

$$\frac{d}{dt}H = \sum_i \dot{p}_i \dot{q}_i + p_i \ddot{q}_i - \frac{d}{dt}\mathcal{L}(q_i, \dot{q}_i, t) \tag{7.18}$$

The time derivative of the Lagrangian, required in Eq. (7.18), can be obtained as follows:

$$\frac{d}{dt}\mathcal{L}(q_i,\dot{q}_i,t)=\frac{\partial\mathcal{L}(q_i,\dot{q}_i,t)}{\partial q_i}\dot{q}_i+\frac{\partial\mathcal{L}(q_i,\dot{q}_i,t)}{\partial\dot{q}_i}\ddot{q}_i+\frac{d}{dt}\mathcal{L}(q_i,\dot{q}_i,t) \quad (7.19)$$

Now, recalling some of the Lagrangian mechanics discussed above, we have already found that:

$$\begin{aligned}\frac{\partial\mathcal{L}(q_i,\dot{q}_i,t)}{\partial q_i}&=\dot{p}_i\\ \frac{\partial\mathcal{L}(q_i,\dot{q}_i,t)}{\partial\dot{q}_i}&=p_i\end{aligned} \quad (7.20)$$

where the first line is a direct consequence of the Euler–Lagrange equation, and the second line is the definition of the canonical momentum. Hence, we can conclude:

$$\frac{d}{dt}\mathcal{L}(q_i,\dot{q}_i,t)=\dot{p}_i\dot{q}_i+p_i\ddot{q}_i+\frac{d}{dt}\mathcal{L}(q_i,\dot{q}_i,t) \quad (7.21)$$

The result in Eq. (7.21) can then be inserted into Eq. (7.18) to obtain the following neat equation:

CONSERVATION OF ENERGY

$$\frac{d}{dt}H=-\frac{d}{dt}\mathcal{L}(q_i,\dot{q}_i,t) \quad (7.22)$$

Eq. (7.22) is a statement for the conservation of total energy that reads as: *the total energy of a system is conserved if the Lagrangian has no explicit time dependence other than the one through position and velocity.* Similarly, Eq. (7.21) constitutes another important statement that reads as: *the Lagrangian is not a conserved quantity because both position and velocity are always time dependent.*

CHAPTER 8

Poisson Brackets

IN THIS CHAPTER I shall discuss what I personally find the most amazing piece of classical mechanics and to do so I begin with a very general statement: *in classical mechanics we deal with functions of position and velocity or, in the Hamiltonian approach, with functions of position and momentum*. Let me indicate those functions with the generic notation $f(q_i, p_i)$. What is the time derivative of the function $f(q_i, p_i)$? Most literally, it can be evaluated as:

$$\dot{f}(q_i, p_i) = \sum_i \frac{\partial f(q_i, p_i)}{\partial q_i} \dot{q}_i + \frac{\partial f(q_i, p_i)}{\partial p_i} \dot{p}_i \tag{8.1}$$

that is, it is the result of the sum of the partial derivatives with respect to every position and momentum coordinates. Recalling Hamilton's equations (see Eq. (7.8)), Eq. (8.1) can be expanded in:

$$\dot{f}(q_i, p_i) = \sum_i \frac{\partial f(q_i, p_i)}{\partial q_i} \frac{\partial H(q_i, p_i)}{\partial p_i} - \frac{\partial f(q_i, p_i)}{\partial p_i} \frac{\partial H(q_i, p_i)}{\partial q_i} \tag{8.2}$$

Its right-hand-side can be isolated and abbreviated as:

$$\{f(q_i, p_i), H(q_i, p_i)\} = \sum_i \frac{\partial f(q_i, p_i)}{\partial q_i} \frac{\partial H(q_i, p_i)}{\partial p_i} - \frac{\partial f(q_i, p_i)}{\partial p_i} \frac{\partial H(q_i, p_i)}{\partial q_i} \tag{8.3}$$

DOI: 10.1201/9781003459781-8

The notation in curly brackets on the left-hand-side of this equation takes the name of *Poisson bracket* after the work of Siméon Poisson who first proposed its definition. There is an enormous power in this operation and a lot of elegance in its form! Combining Eqs. (8.2) and (8.3) we get:

POISSON BRACKET

$$\dot{f}(q_i,p_i) = \{f(q_i,p_i), H(q_i,p_i)\} \tag{8.4}$$

Let me read out loud what Eq. (8.4) means: *the time derivative of a function of position and momentum is simply the Poisson bracket between the function and the Hamiltonian.* Essentially, the core of classical mechanics in one single equation, mathematical elegance at its best! There is a reason why I sound so excited, and I shall make it clear soon.

But before demonstrating the power of the Poisson bracket through some practical examples, let me enumerate a few of its most important properties. Indicating the functions required in the Poisson bracket generically as f and g (I dropped their variables to simplify the notation), the generic Poisson bracket is:

$$\{f,g\} = \sum_i \frac{\partial f}{\partial q_i}\frac{\partial g}{\partial p_i} - \frac{\partial f}{\partial p_i}\frac{\partial g}{\partial q_i} \tag{8.5}$$

Here some of its properties:

1. The Poisson bracket is a linear operation, i.e.:

$$\{kf,g\} = k\{f,g\} \tag{8.6}$$

with k a constant. And:

$$\{h+f,g\} = \{h,g\} + \{f,g\} \tag{8.7}$$

with h a function.

2. The Poisson bracket of a product of two functions is:

$$\{hf,g\} = \{h,g\}f + h\{f,g\} \tag{8.8}$$

3. The Poisson bracket is antisymmetric with respect to the exchange of the two functions, i.e.:

$$\{f,g\} = -\{g,f\} \tag{8.9}$$

4. The Poisson bracket of a function with itself is always zero, i.e.:

$$\{f,f\} = 0 \tag{8.10}$$

If you are thinking that these properties look familiar to you, well you are right! Have a look at the properties of the *commutator* introduced in Chapter 2 (Eqs. (2.24)–(2.28)) to find out the amazing similitude between the commutator (a fundamental tool in quantum mechanics) and the Poisson bracket (a fundamental tool in classical mechanics). I shall come back to this later, but first let me go over a few important applications where the full potential of the Poisson bracket becomes evident:

- The Poisson bracket between a function f and the canonical momentum is readily calculated using Eq. (8.5) and setting $g = p$ to obtain:

$$\begin{aligned} \{f,p\} &= \frac{\partial f}{\partial q}\frac{\partial p}{\partial p} - \frac{\partial f}{\partial p}\frac{\partial p}{\partial q} \\ &= \frac{\partial f}{\partial q} \end{aligned} \tag{8.11}$$

  where I have used the fact that $\frac{\partial p}{\partial q} = 0$ because position and momentum are two separate coordinates. Reading the result from right to left we have:

GENERATOR OF TRANSLATION IN THE POSITION SPACE

$$\frac{\partial f}{\partial q} = \{f,p\} \tag{8.12}$$

which reads: *the Poisson bracket of a function with the momentum returns the partial derivative of such function with respect to the conjugated coordinate.*

Moreover, considering the very definition of a derivative in the form:

$$\frac{\partial f(q,p)}{\partial q} = \frac{f(q+\delta,p) - f(q,p)}{\delta} \tag{8.13}$$

and rearranging it as:

$$f(q+\delta,p) = f(q,p) + \delta\{f(q,p),p\} \tag{8.14}$$

we realize that the true meaning of Eq. (8.14) is that the spatial derivative, taken with respect to the coordinate q, *generates* a translation along such a coordinate. Now, since the spatial derivative in question is equal to the Poisson bracket between the function and the canonical momentum, then we can conclude that: *the canonical momentum is a generator of translations along the corresponding generalized coordinate*.

- The Poisson bracket between the function f and the position coordinate is readily calculated from Eq. (8.5) by setting $g = q$ to obtain:

$$\begin{aligned} \{f,q\} &= \frac{\partial f}{\partial q}\frac{\partial q}{\partial p} - \frac{\partial f}{\partial p}\frac{\partial q}{\partial q} \\ &= -\frac{\partial f}{\partial p} \end{aligned} \tag{8.15}$$

  Following the same reasoning as in the previous point we conclude that *the Poisson bracket between a function and the position coordinate returns (minus) the partial derivative of the function with respect to the momentum*. Hence, *the position is a generator of translations in the phase space*.

GENERATOR OF TRANSLATION IN THE PHASE SPACE

$$\frac{\partial f}{\partial p} = -\{f,q\} \tag{8.16}$$

- The notion of angular momentum has been introduced in Eq. (5.18). It coincides with the canonical momentum conjugated to the angular velocity of an object rotating about an axis according to:

$$L = \frac{\partial \mathcal{L}}{\partial \dot{\theta}} \tag{8.17}$$

The vector cross product featured in Eq. (5.18) can be worked out to retrieve the three individual components of the angular momentum as:

$$\begin{aligned} L_x &= yp_z - zp_y \\ L_y &= zp_x - xp_z \\ L_z &= xp_y - yp_x \end{aligned} \tag{8.18}$$

The Poisson brackets between each of the three components of position and the z-component of the angular momentum result in:

$$\begin{aligned} \{x, L_z\} &= -y \\ \{y, L_z\} &= x \\ \{z, L_z\} &= 0 \end{aligned} \tag{8.19}$$

The result in Eq. (8.19) can be read a rotation about the z-axis and generalized into the statement: *the Poisson bracket involving the angular momentum about an axis returns a rotation about such axis.* Hence, *the angular momentum is a generator of spatial rotations.*

- The Poisson bracket between position and momentum is calculated from:

$$\{q, p\} = \frac{\partial q}{\partial q}\frac{\partial p}{\partial p} - \frac{\partial q}{\partial p}\frac{\partial p}{\partial q} \tag{8.20}$$

However, because of the following equalities:

$$\begin{aligned} \frac{\partial p}{\partial p} &= \frac{\partial q}{\partial q} = 1 \\ \frac{\partial q}{\partial p} &= \frac{\partial p}{\partial q} = 0 \end{aligned} \tag{8.21}$$

Eq. (8.20) can be further simplified into:

$$\{q, p\} = 1 \tag{8.22}$$

As explained in Chapter 10, this is a crucial result that has a very famous counterpart in quantum mechanics.

- Finally, and crucially to the scope of this book, I evaluate the Poisson bracket between a generic function f and the Hamiltonian. This is calculated from Eq. (8.5) by setting $g = H$ to obtain:

$$\{f,H\} = \frac{\partial f}{\partial q}\frac{\partial H}{\partial p} - \frac{\partial f}{\partial p}\frac{\partial H}{\partial q} \tag{8.23}$$

Recalling Hamilton's equations (see Eq. (7.8)), the last equality can be rewritten as:

$$\{f,H\} = \frac{\partial f}{\partial q}\dot{q} + \frac{\partial f}{\partial p}\dot{p} \tag{8.24}$$

You may note that the right-hand-side of Eq. (8.24) coincides with the time derivative of the function f and therefore:

GENERATOR OF TIME EVOLUTION

$$\frac{df}{dt} = \{f,H\} \tag{8.25}$$

The meaning of Eq. (8.25) is that *the Poisson bracket between a function and the Hamiltonian returns the time derivative of such function*. Hence, *the Hamiltonian is a generator of time translations.*

Before moving on to read the next chapter, I would like the reader to look once more at Eq. (8.25) and compare it to Eq. (2.29); have some thinking while I briefly discuss about conservations law before to come back to this comparison in a final chapter.

CHAPTER 9

Conservation Laws

IN PREVIOUS CHAPTERS, AND within different representations of classical mechanics, I have introduced the conservation laws for linear momentum, angular momentum, and total energy. Physicists have figured out a common pattern that underpins all such conservation laws and managed to rationalize it in terms of *change of coordinates*, *symmetry of the Lagrangian*, and the concept of *generators of motion*. Since the ground has been laid, I shall now do my best to present these conservation laws in a hopefully easy-to-read way.

I have already discussed a bit about coordinates, making extensive use of Cartesian as well as generalized coordinates. I shall now discuss briefly about changes of coordinates. At its very core, such discussion assumes that *there is no special place in the Universe*. Hence, the laws of physics must be *invariant* under a change of coordinates. This means that there is nothing like absolute coordinates and absolute reference frames, since those tools are simply a construct that scientists have proposed in order to facilitate our understanding of the laws of Nature. Because of this, the laws of physics (which exists before us) must not change depending on our choice of coordinates. In particular, *the laws of motion must not change under a translation, nor under a rotation of coordinates, nor under a translation in time*. The latter is true because *there is no such a thing such as absolute time*; time is rather a coordinate, likewise any of the spatial coordinates encountered before.

A change of coordinates can be seen as generated by the buildup of whatever many infinitesimal changes in each coordinate. If I indicate with

DOI: 10.1201/9781003459781-9

ε the infinitesimal step change, then any generic transformation of coordinates can be parametrized as:

$$q_i' = q_i + c_i\left(q_i\right)\varepsilon \tag{9.1}$$

where q_i' is the transformed coordinate, q_i is its original value and $c_i\left(q_i\right)$ is a coefficient regulating such change. $c_i\left(q_i\right)$ is, in general, a function of the whole set of generalized coordinates, i.e., the transformation may depend upon where in the configuration space it happens.

Let me consider a *spatial translation* first; the term spatial translation indicates a coordinate transformation that shifts coordinates in space by adding a constant amount to it. A very basic but important spatial translation is obtained when all coordinates are shifted by the same amount, no matter where the object is in the configuration space. This is described by the transformation:

$$q_i' = q_i + \varepsilon \tag{9.2}$$

where all $c_i\left(q_i\right) = 1$. This coordinates transformation can be visualized in a simple bidimensional case as shown in Figure 9.1.

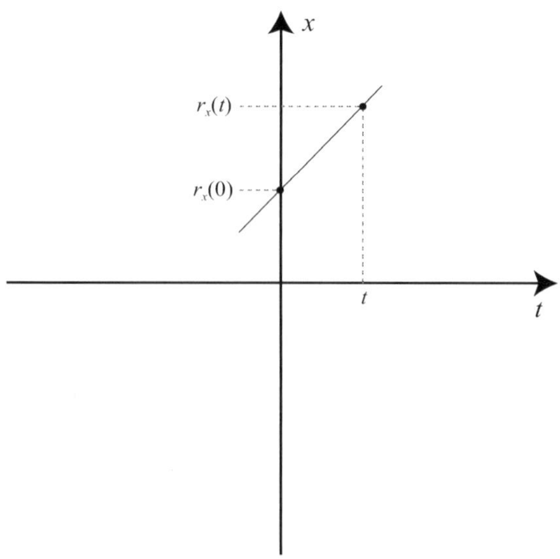

FIGURE 9.1 A spatial translation transformation in two dimensions of the coordinates q_1 and q_2 by an equal amount ε.

Note that it is perfectly allowed that a translational transformation of coordinates is such that some coordinates are shifted by an amount and some other coordinates are shifted by a different amount, such as in the three-dimensional example below:

$$\begin{aligned} q'_x &= q_x + a\varepsilon \\ q'_y &= q_y + b\varepsilon \\ q'_z &= q_z \end{aligned} \tag{9.3}$$

In such case, the coefficients occurring in Eq. (9.1) would be: $c_x\left(q_x\right) = a$, $c_y\left(q_y\right) = b$ and $c_z\left(q_z\right) = 0$.

Coordinates can also be rotated. For example, we may think of the transformation occurring when both the x and y coordinates are rotated about the z-axis as shown in Figure 9.2. The transformation rules to be used in the case of rotations are usually more complicated than those for translations, nevertheless they can be figured out with the use of trigonometry and polar coordinates.

As a manageable example, I can show the procedure that leads to the coordinate transformations rules for the case reported in Figure 9.2. Firstly, the coordinates x and y of the point P in the original reference frame are written in terms of the polar coordinates R and θ as:

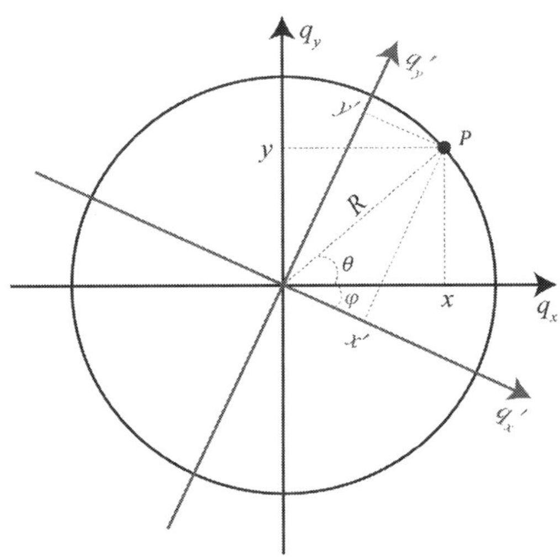

FIGURE 9.2 A spatial rotation transformation in two dimensions of the coordinates q_x and q_y by an angle φ about the q_z axis.

$$\begin{aligned} x &= R\cos(\theta) \\ y &= R\sin(\theta) \\ z &= z \end{aligned} \tag{9.4}$$

Similarly, the coordinates x' and y' of the point P as seen from the rotated frame are derived as:

$$\begin{aligned} x' &= R\cos(\theta+\varphi) = R\cos(\theta)\cos(\varphi) - R\sin(\theta)\sin(\varphi) \\ y' &= R\sin(\theta+\varphi) = R\sin(\theta)\cos(\varphi) + R\cos(\theta)\sin(\varphi) \\ z' &= z \end{aligned} \tag{9.5}$$

where I have used trigonometry to expand the sine and cosine of a sum of angles. The coordinates in Eq. (9.5) can be further manipulated to be expressed in terms of the original coordinates with the use of the definitions given in Eq. (9.4) to obtain:

$$\begin{aligned} x' &= x\cos(\varphi) - y\sin(\varphi) \\ y' &= y\cos(\varphi) + x\sin(\varphi) \\ z' &= z \end{aligned} \tag{9.6}$$

Consider now an infinitesimal rotation (i.e. set $\varphi = \varepsilon$), and use a Taylor expansion, truncated to first order, to rewrite sine and cosine occurring in Eq. (9.6) as:

$$\begin{aligned} \cos(\varphi) &= \cos(\varepsilon) \sim 1 \\ \sin(\varphi) &= \sin(\varepsilon) \sim \varepsilon \end{aligned} \tag{9.7}$$

The coordinates in Eq. (9.6) will then become:

$$\begin{aligned} x' &= x - y\,\varepsilon \\ y' &= y + x\,\varepsilon \\ z' &= z \end{aligned} \tag{9.8}$$

where, by comparison with Eq. (9.1), we find $c_x(q_x) = -y$, $c_y(q_y) = x$ and $c_z(q_z) = 0$.

At the beginning of the chapter, I mentioned we are going to make use of symmetries. A symmetry is *an infinitesimal transformation of coordinates that does not change the Lagrangian*. Thus, to verify whether a transformation is a symmetry, we need to write down the Lagrangian (the way discussed in Chapter 5) and check whether or not it changes under a proposed transformation. Let me assume we have found a symmetry for an object for which we know the Lagrangian. For what said, this means that we have found a coordinate transformation that does not change the Lagrangian. Let me indicate such transformation as:

$$q_i' = q_i + s_i(q_i)\varepsilon \tag{9.9}$$

where I used $s_i(q_i)$, rather than $c_i(q_i)$, to highlight that this transformation has the features of a symmetry. For the very definition of a symmetry, I expect to find out that the total variation of the Lagrangian (what I would call $\delta\mathcal{L}$, i.e., the sum of all derivatives of the Lagrangian with respect to every single variable) is zero. In mathematics, such variation is formally written as:

$$\underset{=0}{\delta\mathcal{L}(q_i,\dot{q}_i,t)} = \sum_i \frac{\partial\mathcal{L}(q_i,\dot{q}_i,t)}{\partial q_i}\delta q_i + \frac{\partial\mathcal{L}(q_i,\dot{q}_i,t)}{\partial \dot{q}_i}\delta \dot{q}_i + \frac{d}{dt}\mathcal{L}(q_i,\dot{q}_i,t)\delta t \tag{9.10}$$

Recalling the results of Eq. (7.20):

$$\frac{\partial\mathcal{L}(q_i,\dot{q}_i,t)}{\partial q_i} = \dot{p}_i$$

$$\frac{\partial\mathcal{L}(q_i,\dot{q}_i,t)}{\partial \dot{q}_i} = p_i$$

the total variation of the Lagrangian can be rewritten as:

$$\underset{=0}{\delta\mathcal{L}(q_i,\dot{q}_i,t)} = \sum_i (\dot{p}_i\delta q_i + p_i\delta\dot{q}_i) + \frac{d}{dt}\mathcal{L}(q_i,\dot{q}_i,t)\delta t \tag{9.11}$$

which, using the product rule of differentiation, simplifies to:

$$\underset{=0}{\delta\mathcal{L}(q_i,\dot{q}_i,t)} = \frac{d}{dt}\sum_i p_i\delta q_i + \frac{d}{dt}\mathcal{L}(q_i,\dot{q}_i,t)\delta t \tag{9.12}$$

Concentrate your attention on the second term on the right-hand-side in the first line of Eq. (9.12): if the Lagrangian has no explicit time dependence (with this I mean, once again, no direct dependence on time other than that due to the time dependence of its variables q_i and $\dot{q}_i$), this second term is equal to zero and the system is said to be *time-translation invariant.* When this occurs, i.e., for time-translation invariant systems, the condition under which a transformation of coordinates is a symmetry reads:

$$\frac{d}{dt}\sum_i p_i \delta q_i = 0 \tag{9.13}$$

The variation of the coordinates, δq_i, featuring in Eq. (9.13), can be directly obtained from Eq. (9.9) as:

$$\begin{aligned} \delta q_i &= q_i' - q_i \\ &= s_i\left(q_i\right)\varepsilon \end{aligned} \tag{9.14}$$

We now have a compact and extremely powerful equation that provides the exact quantity to be conserved under a specific symmetry:

CONSERVED QUANTITY UNDER A SYMMETRY

$$\frac{d}{dt}\sum_i p_i s_i\left(q_i\right) = 0 \tag{9.15}$$

where the constant factor ε has been dropped in going from Eq. (9.14) to Eq. (9.15) because it is only a multiplicative constant. The meaning of Eq. (9.15) is that, providing that a coordinate transformation parametrization is known, one can evaluate whether such transformation is, or is not, a symmetry of the system. And there is more to it! Equation (9.15) is also a conservation law since it states that a certain quantity (the summation in Eq. 9.15) does not vary with time. Hence, we can conclude that: *having a symmetry implies that a certain physical quantity is conserved; the conserved quantity being* $\sum_i p_i s_i\left(q_i\right)$.

For the sake of completeness and rigour, I also need to state that, by using the Euler–Lagrange equation in rewriting Eq. (9.10), I have implicitly assumed that the system can be described by Lagrangian mechanics, which, in turn, requires the assumption of the stationary action principle. Hence, the connection between symmetry and conservation laws discussed in this chapter is only valid for Lagrangian that are derived from such a principle. As done before, I shall now go through a few examples with the aim of clarifying the content of this chapter.

- Consider an object described by the following Lagrangian:

$$\mathcal{L} = \frac{1}{2} m\left(\dot{q}_1^2 + \dot{q}_2^2\right) - V\left(q_1 + q_2\right) \tag{9.16}$$

  and say we want to check whether the following translational transformation:

$$\begin{aligned} q_1' &= q_1 + \varepsilon \\ q_2' &= q_2 + \varepsilon \end{aligned} \tag{9.17}$$

  is a symmetry of the system. Given the fact that ε is a constant, the derivatives of the coordinates are quite simple to work out as:

$$\begin{aligned} \dot{q}_1' &= \dot{q}_1 \\ \dot{q}_2' &= \dot{q}_2 \end{aligned} \tag{9.18}$$

  We can conclude that the kinetic energy term in the Lagrangian is invariant under the proposed transformation. However, it is also quite simple to see how the potential energy term changes into something else when both coordinates assume the values q_1' and q_2' given in Eq. (9.17). Hence, we ultimately conclude that the transformation is not a symmetry.

- Consider, in a second example, the same Lagrangian and the following translational transformation:

$$\begin{aligned} q_1' &= q_1 + \varepsilon \\ q_2' &= q_2 - \varepsilon \end{aligned} \tag{9.19}$$

  The derivatives of the coordinates do coincide with the ones shown in Eq. (9.18). Moreover, since the potential energy in Eq. (9.16) depends upon the sum of the two generalized coordinates, the whole Lagrangian does not change in this case. The transformation written in Eq. (9.19) is therefore a symmetry and we can be sure that the quantity given in Eq. (9.15) is a conserved quantity for the system. This quantity evaluates to:

$$
\begin{aligned}
Q &= \sum_i p_i s_i(q_i) \\
&= p_1 - p_2
\end{aligned}
\tag{9.20}
$$

That is, the difference between the two momenta is conserved.

- In a third and final example, I consider a system described by the following Lagrangian:

$$
\mathcal{L} = \frac{1}{2} m\left(\dot{q}_1^2 + \dot{q}_2^2\right) - V\left(q_1^2 + q_2^2\right) \tag{9.21}
$$

to check whether the following rotational transformation (previously encountered in Eq. (9.8)):

$$
\begin{aligned}
q_1' &= q_1 - q_2\varepsilon \\
q_2' &= q_2 + q_1\varepsilon
\end{aligned}
\tag{9.22}
$$

is a symmetry of the system. The derivatives of the two coordinates are readily calculated as:

$$
\begin{aligned}
\dot{q}_1' &= \dot{q}_1 - \dot{q}_2\varepsilon \\
\dot{q}_2' &= \dot{q}_2 + \dot{q}_1\varepsilon
\end{aligned}
\tag{9.23}
$$

The sum of their squares, needed in the kinetic energy term, does not change to a first-order approximation in ε, as can be seen by writing:

$$
\begin{aligned}
\dot{q}_1'^2 + \dot{q}_2'^2 &= \dot{q}_1^2 + \dot{q}_2^2 + \dot{q}_1^2\varepsilon^2 + \dot{q}_2^2\varepsilon^2 \\
&\cong \dot{q}_1^2 + \dot{q}_2^2
\end{aligned}
\tag{9.24}
$$

A similar argument can be made for the sum of the squared coordinates occurring in the potential energy term. The transformation in Eq. (9.23) is therefore a symmetry. The system must therefore have a conserved quantity, and this is given by:

$$
\begin{aligned}
Q &= \sum_i p_i s_i(q_i) \\
&= -q_2 p_1 + p_2 q_1
\end{aligned}
\tag{9.25}
$$

where I have used $s_1(q_1) = -q_2$ and $s_2(q_2) = q_1$ as appearing in Eq. (9.22). The quantity Q in Eq. (9.25) coincides with the angular momentum along the z axis (make the substitution $q_1 = x$, $q_2 = y$, $p_1 = p_x$ and $p_2 = p_y$ and compare it with Eq. (8.18)).

Let me summarize what was discussed in this chapter in a single statement: *if the Lagrangian of the system is invariant under a given transformation of coordinates then there is a quantity that is conserved and, hence, a conservation law exists.*

In the usual quest for mathematic elegance, more can be added when the Poisson bracket is taken into consideration. In Chapter 8, we have seen how some special functions (there generically indicated as $g(q,p)$) can generate transformations such as spatial translations, spatial rotations, or time translations. If those transformations are a symmetry of the system, then the energy of the system is not changed by such a transformation. This fact can also be expressed as: *for a symmetry, the total change in the Hamiltonian due to the transformations is null.* This condition can be written in terms of a Poisson bracket as:

$$\begin{aligned}\delta H &= \{H(q,p), g(q,p)\} \\ &= 0\end{aligned} \tag{9.26}$$

However, since Poisson brackets are antisymmetric (see Eq. (8.9)), the last equation can be also written as:

$$\{g(q,p), H(q,p)\} = 0 \tag{9.27}$$

and we already know that the Poisson bracket in Eq. (9.27) returns the time derivative of the function $g(q,p)$ (see Chapter 8, Eq. (8.25)). I can therefore conclude that: *when $g(q,p)$ is a conserved quantity (as verified by satisfying* Eq. (9.27)), *then it is also a symmetry for the system; hence, $g(q,p)$ generates a transformation under which a given quantity is conserved.*

CHAPTER 10

Quantum Mechanics (Conception)

As announced, I start this final chapter by explicitly comparing Eq. (2.29) of quantum mechanics to Eq. (8.25) of classical mechanics (Poisson's formulation), namely, by comparing:

$$\frac{d}{dt}\langle\hat{\mathrm{O}}\rangle = -\frac{i}{\hbar}\left[\hat{\mathrm{O}},\hat{\mathrm{H}}\right]$$

with

$$\frac{df}{dt} = \{f,H\}$$

There is an astonishing similarity between those two equations, and this is perhaps expected since the two theories describe the same phenomena, although within different regimes. In quantum mechanics (see Chapter 2) a physical observable, say the linear momentum, energy, or angular momentum of an object is represented by an operator. The theoretically predicted value of these observables, to be matched to the result from a statistically meaningful number of measurements of such observable, is given by the expectation value of the operator, $\langle\hat{\mathrm{O}}\rangle$, calculated as in Eq. (2.13). The time dependence of the expectation value is proportional to the commutator between the operator and the Hamiltonian for the system.

DOI: 10.1201/9781003459781-10

In classical mechanics, position, linear momentum, *etc.*, are functions of time. Their time dependence is proportional to the Poisson bracket between the function representing the observable and the Hamiltonian. In other words, the equation describing the dynamics of objects in classical mechanics, can be *obtained* from the more general quantum mechanics by replacing the commutator featuring in Eq. (2.29) with a Poisson bracket multiplied by the reduced Planck constant and the imaginary unit, as in:

$$\left[..,..\right] \leftrightarrow i\hbar\left\{..,..\right\} \tag{10.1}$$

Above, I have used italic fonts for the word *obtained* to warn the reader that this is not a formal derivation but a mere comparison. Historically, classical mechanics was proposed much earlier than quantum mechanics and it was Paul Dirac, one of the pioneers of quantum mechanics, to note that equations from classical mechanics could be *quantized* by replacing Poisson brackets with commutators (and the multiplicative factors as in Eq. (10.1)). Note the centrality of the Hamiltonian that, as known already from classical mechanics, is the generator of time translations (see Chapter 8). I shall come back to discuss about this comparison later, but first I want to introduce the operators for *position* and *linear momentum*, the two central quantities we have repeatedly used across all formulations of classical mechanics. For the sake of simplicity, I limit the discussion to a one-dimensional problem where the object can only move along the x-direction, and therefore, it is described by a wavefunction which is a function of the sole spatial variable x.

The operator associated with the position of an object along the x-axis is defined as:

POSITION OPERATOR

$$\hat{x} \Rightarrow x \tag{10.2}$$

meaning that it acts on the wavefunction by taking the product between the position variable and the wavefunction as in:

$$\hat{x}\Psi(x) \Rightarrow x\Psi(x) \tag{10.3}$$

As explained in Chapter 2, the most general action of an operator on a wavefunction is to generate another wavefunction. However, one can write an eigenequation similar to that in Eq. (2.4) and solve it to find eigenvalues and eigenfunctions of a given operator. For the position operator in Eq. (10.3), this would read as:

$$\hat{x}\psi(x) = x_0\psi(x) \tag{10.4}$$

which, using Eq. (10.2), becomes:

$$x\psi(x) = x_0\psi(x) \tag{10.5}$$

where, $\psi(x)$ is now an eigenfunction of the position operator $\hat{x}$ with associated eigenvalue x_0. Note that, as done in Chapter 2, I have used a different symbol to distinguish an eigenfunction, $\psi(x)$, from the generic wavefunction, $\Psi(x)$. Recall that if the object under investigation happens to be described by the actual wavefunction $\psi(x)$ (which is an eigenfunction of the position operator) then the expectation value of its position will be x_0, i.e., the object represented by such wavefunction is in x_0. But what about the explicit mathematical form of $\psi(x)$? To find this, I rewrite Eq. (10.5) as:

$$(x - x_0)\psi(x) = 0 \tag{10.6}$$

In this form it is clearer that the equation implies that when x is different from x_0 then $\psi(x) = 0$, or, in other words, that the wavefunction is non-zero only at the specific location x_0 and null in any other. The mathematical function that has such feature is the so-called *Dirac delta*, which is formally written as:

EIGENFUNCTIONS OF THE POSITION OPERATOR

$$\psi(x) = \delta(x - x_0) \tag{10.7}$$

Dirac delta functions are more rigorously defined through the following relation:

$$\int_{-\infty}^{+\infty} \delta(x - x_0) f(x) dx = f(x_0) \tag{10.8}$$

which defines the Dirac delta function as the one returning the value of a function at a specific value of its variable.

To summarize what has been discussed so far: the position operator, $\hat{x}$, acts on a wavefunction by multiplying the variable x by the wavefunction itself. Its eigenvalues, x_0, are the positions along the x-axis taken by the described object (they are infinite in number) and its eigenfunctions are the Dirac delta functions $\delta(x - x_0)$. The position operator is Hermitian, hence, its eigenvalues are real and its eigenfunctions form an orthogonal basis set.

The operator associated with the linear momentum of an object along the x-axis is defined as:

LINEAR MOMENTUM OPERATOR

$$\hat{p}_x \Rightarrow -i\hbar \frac{\partial}{\partial x} \tag{10.9}$$

It acts on the wavefunction by taking its derivative with respect to the x component of the object's position variable (and multiplying the result for $-i\hbar$) as in:

$$\hat{p}_x \Psi(x) \Rightarrow -i\hbar \frac{\partial \Psi(x)}{\partial x} \tag{10.10}$$

As we have done for the position, we can write an eigenequation for the linear momentum operator:

$$\hat{p}_x \psi(x) = p_x \psi(x) \tag{10.11}$$

where $\psi(x)$ now represents the set of eigenfunctions (or eigenvectors) of the linear momentum operator along the x-axis and p_x represents the set of associated eigenvalues (one for each eigenfunction in the set). These eigenvalues correspond to the values of the linear momentum of the object described by a given eigenfunction (see Eq. (2.5)). To find the mathematical form of the eigenfunctions of the linear momentum operator, we write the eigenequation in Eq. (10.11) using the definition in Eq. (10.9) as in:

$$-i\hbar\frac{\partial}{\partial x}\psi(x)=p_x\psi(x) \tag{10.12}$$

This can be rearranged as:

$$\frac{\partial}{\partial x}\psi(x)=i\frac{p_x}{\hbar}\psi(x) \tag{10.13}$$

The differential equation above simply asks for a function (in fact, a whole class of functions) whose derivative is proportional to the function itself. This is a relatively well-known problem in mathematics and the class of functions we are looking for is the class of exponential functions. In the current case, the solution to Eq. (10.13) gives:

EIGENFUNCTIONS OF THE LINEAR MOMENTUM OPERATOR

$$\psi(x)=Ae^{i\frac{p_x}{\hbar}x} \tag{10.14}$$

where A is a constant. This can be easily verified by substituting the result in Eq. (10.14) back into Eq. (10.13) and see that the equality is verified for whatever value of A.

One may now ask why the linear momentum operator has the actual form as given above. The connection between the linear momentum and the spatial derivative is well-known in classical mechanics, and I have discussed about linear momentum as the generator of spatial translations in Chapter 8 (see Eq. (8.14)). The presence of the reduced Planck constants, whose units are J s, is required for unit consistency. Basically, the argument of an exponential is a pure (unitless) number; the units of the linear momentum, from classical mechanics, are kg m s^{-1} and the units of position are in m, hence we need to divide by J s for the term $p_x x$ in the argument of the exponential to be dimensionless. Finally, the presence of the imaginary units is required to satisfy Hermitianity. I have explained in Chapter 2 that physical observables need to be represented by Hermitian operators since these latter have real eigenvalues. The condition for an operator to be Hermitian has been given in Eq. (2.23) and requires that the operator coincides with its adjoint (also known as its Hermitian conjugate). However, to verify whether an operator is Hermitian is more conveniently done by satisfying the following equivalent condition:

$$\int_{-\infty}^{\infty} \phi^*(x)\hat{O}\psi(x)dx = \left(\int_{-\infty}^{\infty} \psi^*(x)\hat{O}\phi(x)dx\right)^* \tag{10.15}$$

This, adapted to the case of the linear momentum operator in Eq. (10.9), becomes:

$$-i\hbar\int_{-\infty}^{\infty} \phi^*(x)\frac{\partial}{\partial x}\psi(x)dx = \left(-i\hbar\int_{-\infty}^{\infty} \psi^*(x)\frac{\partial}{\partial x}\phi(x)dx\right)^* \tag{10.16}$$

To verify whether such an equality holds, I solve the integral on the right-hand-side by part to obtain:

$$\int_{-\infty}^{\infty} \psi^*(x)\frac{\partial}{\partial x}\phi(x)dx = -\int_{-\infty}^{\infty} \phi(x)\frac{\partial}{\partial x}\psi^*(x)dx \tag{10.17}$$

and insert this result into Eq. (10.16) to find:

$$-i\hbar\int_{-\infty}^{\infty} \phi^*(x)\frac{\partial}{\partial x}\psi(x)dx = \left(i\hbar\int_{-\infty}^{\infty} \phi(x)\frac{\partial}{\partial x}\psi^*(x)dx\right)^* \tag{10.18}$$

where the term on the right-hand-side become identical to the one on the left-hand-side of the equality once the conjugate operation is taken. The result in Eq. (10.18) verifies that the linear momentum operator is Hermitian and therefore has real eigenvalues. An opposite conclusion, i.e., the operator would have resulted non-Hermitian, if the imaginary unit was omitted in Eq. (10.9). Note that the same reasoning can be followed to prove that the position operator in Eq. (10.2) is also Hermitian, as expected for an operator representing the position observable.

Now that the operators for both position and linear momentum observables are introduced, we can evaluate the commutator between the two. This is readily made as:

$$\left[\hat{x},\hat{p}_x\right]=\left[x,-i\hbar\frac{\partial}{\partial x}\right]$$
$$=-i\hbar\left(x\frac{\partial}{\partial x}-\frac{\partial}{\partial x}x\right) \tag{10.19}$$

The commutator above, is still an operator, and as such can be applied to a wavefunction, to give:

$$-i\hbar\left(x\frac{\partial}{\partial x}-\frac{\partial}{\partial x}x\right)\Psi(x)=-i\hbar\left(x\frac{\partial\Psi(x)}{\partial x}-\frac{\partial\left(x\Psi(x)\right)}{\partial x}\right)$$
$$=-i\hbar\left(x\frac{\partial\Psi(x)}{\partial x}-\Psi(x)-x\frac{\partial\Psi(x)}{\partial x}\right) \tag{10.20}$$
$$=i\hbar\Psi(x)$$

Hence, the result:

COMMUTATOR BETWEEN POSITION AND MOMENTUM

$$\left[\hat{x},\hat{p}_x\right]=i\hbar \tag{10.21}$$

This is a remarkable result for many reasons. Firstly, consider the classical counterpart of Eq. (10.21), i.e., the Poisson bracket calculated in Eq. (8.22), which I report here for convenience:

$$\{q,p\}=1$$

At the beginning of this chapter (see Eq. (10.1)), I have briefly discussed about Dirac's quantization method consisting in identifying the commutator used in quantum mechanics with the Poisson bracket used in classical mechanics multiplied by $i\hbar$. This statement works perfectly also with Eq. (10.21) that can be written as:

$$\left[\hat{x},\hat{p}_x\right]=i\hbar\{q,p\} \tag{10.22}$$

given the fact that the Poisson bracket on the right-hand-side evaluates to unity.

The real importance of Eq. (10.21), however, lies in a more fundamental concept, the discussion of which requires some new insights into commutators between operators. Basically, it can be demonstrated that operators that commute, i.e., operators that satisfy the following relationship:

$$\left[\hat{O}_1, \hat{O}_2\right] = 0 \tag{10.23}$$

share a common set of eigenfunctions. Although not straightforward to understand, this will ultimately lead to the fact that the two physical observables associated with the two operators $\hat{O}_1$ and $\hat{O}_2$ can be simultaneously measured in an experiment. Now, this latter conclusion may seem trivial since we may expect to be always able to simultaneously measure any two (or more) quantities of interest. However, such an ability is only admitted in classical mechanics, hence, it is only available when measuring properties of macroscopic objects. For microscopic object, whose behaviour is interpreted through quantum mechanics, the ability to measure two properties at the same time is not trivially true but rather, it depends on the value of the commutator between the operators representing the two physical properties under scrutiny. This surely adds a new item to the list of *oddities* encountered by students approaching quantum mechanics. Let me fix this crucial result of quantum mechanics with a more precise statement:

> *only physical observables whose associated operators commute can be measured simultaneously and with an arbitrary precision.*

Please note the condition *with an arbitrary precision* concluding the statement above. The reason for it lies in another quite fundamental point of quantum mechanics known as the *Heisenberg's uncertainty principle.* Basically, using the famous *Cauchy–Schwarz* inequality, it is possible to derive a rule that gives a minimum threshold for the product of the uncertainty in measuring any two physical observables. The *Cauchy–Schwarz* inequality, extended to the space of wavefunctions (the Hilbert space), becomes:

$$\Delta O_1 \, \Delta O_2 = \frac{1}{2} |\langle \psi | [\hat{O}_1, \hat{O}_2] | \psi \rangle | \tag{10.24}$$

where ΔO_1 and ΔO_2 are the uncertainties in measuring the observables O_1 and O_2 associated to the operators $\hat{O}_1$ and $\hat{O}_2$, respectively. Equation (10.24) makes clear the meaning of the statement above: two observables can be measured with arbitrary precision (infinite precision) only if the product of their measurements' uncertainty is null. This can only happen when the commutator between the two associated operators is null.

I can now come back to the importance of the result in Eq. (10.21): in quantum mechanics, the operators representing the position and linear momentum of a microscopic object do not commute. Hence, we cannot measure the object's position and linear momentum, simultaneously and with an arbitrary precision. Moreover, given the result in Eq. (10.24), we can calculate a threshold for the product of the uncertainties in measuring those two properties as:

HEISENBERG'S UNCERTAINTY PRINCIPLE

$$\Delta x \, \Delta p_x = \frac{1}{2} || \langle \psi \, | [\hat{x}, \hat{p}_x] | \psi \rangle || = \frac{\hbar}{2} \tag{10.25}$$

And, if you are thinking on the lines: fine, but the reduced Planck's constant is really a small number of the order of 10^{-34} J s, so who cares? Then, I suggest you consider the value of the Planck's constant expressed in *microscopic units* to find out that $\hbar = 0.39$ g mol^{-1} μm^{-2} μs^{-1} which is clearly not such a small number for what matter to particles, atoms, and molecules!

I hope I managed to show in this chapter how quantum mechanics is related to known concepts of classical mechanics (the Poisson brackets, the Hamiltonian, etc) so that it could be intended as conceived from it. At the same time, I imagine you have realised by now how far from classical mechanics its predictions are! To read more about all this check the books cited in Further Reading.

Ignorance, Scientific Method, and the Tools of Logic

As mentioned earlier in the Introduction, it is my intention to conclude this book exploring some broader concepts that I believe are essential to every student pursuing a scientific degree.

I initiate this discussion with a philosophical inquiry: *Where does scientific research originate*? It is a deliberately broad and open-ended question. What is the fundamental basis upon which we must stand before embarking on scientific pursuits? Is there something we should grasp before delving into this captivating discipline? A discerning reader might remark, "Isn't the pursuit and expansion of knowledge the central aim of science?" Yes, indeed, I would affirm, but it all begins with *ignorance*. Yes, you heard me correctly — it commences from the very antithesis of knowledge. Peculiar, isn't it? To be a scientist, one must embrace ignorance. "Ignorant?" the reader may inquire, "Alright, but only initially, right?" Not at all. A true scientist remains ignorant even at the culmination of their research. Otherwise, it cannot be deemed as scientific inquiry! Ignorance serves as the starting point from which we strive for knowledge. Scientists must embrace their ignorance. I, as a scientist, am ignorant. In truth, all my colleagues are ignorant, and such are many individuals you know. We

DOI: 10.1201/9781003459781-11

are aware that we are bound to be ignorant about certain matters simply because *we cannot know everything.* I could develop this further, but I prefer to pause here and emphasize that the ignorance I speak of is not a derogatory term. On the contrary, scientists must take pride in acknowledging their ignorance.

Before to go any further, I feel compelled to issue a warning: I am not a formally trained philosopher, despite my fascination with philosophy. Thus, I must caution that my forthcoming discussion may present an oversimplified perspective on this subject. However, before digging deeper into the matter, allow me to express my belief that science and philosophy are intricately interconnected to the extent that it is likely impossible to engage in scientific endeavours without the influence of philosophy. It is worth noting that the reverse is possible: doing philosophy without science has been observed throughout history for countless centuries. With that said, I define *doing philosophy* in this context as *employing logic* in the pursuit of philosophical inquiry.

Ignorance is associated either with the lack of information (lack of *data*[1]) or with the lack of understanding of information (lack of a *theory*[1]). There are (at least in my non-expert oversimplified vision) essentially three categories of ignorance that need to be discussed in this chapter: *willful ignorance, unconscious ignorance,* and *conscious ignorance.*

Willful ignorance refers to the conscious refuse to pay attention to information. Note that I wanted to explicitly use the word *conscious* in this definition and yet imply that this form of ignorance is different from *conscious ignorance.* As the definition may suggest, this form of ignorance is often present in law disputes where a judge tries to establish whether information has been deliberately neglected (and therefore willful ignorance built) to gain some advantage. In this category, ignorance is a matter of choice. Things like *fake news, false beliefs*, or *misinformation*, all fall within this category.

Unconscious ignorance refers to the lack of information or the lack of understanding without a conscious recognition of such deficit. In this case, ignorance is not a matter of choice.

I left *conscious ignorance* for last since, as you may imagine, it is the most relevant to this book. This form of ignorance refers to the lack of information or the lack of understanding of which we come to be aware of. As such, this is the ignorance at the basis of science and more specifically at the basis of scientific research. We all have heard at least once that the process of scientific research starts with a good question. But it is the *conscious ignorance* of something that stimulates such a good question. It is in this sense that ignorance comes before everything else in science.

Allow me to provide an example that can help clarify this distinction and potentially alleviate any concerns from specialists in the field regarding my intended meaning with these terms. Let's consider the concept of a black hole (a rather ambitious request!). The idea of black holes gradually took shape over the past 250 years. It was first introduced by John Michell in 1783, but it wasn't until the 1960s that Roger Penrose discussed their possible formation and stability. Only in the 1990s did Genzel and Ghez gather evidence that could indicate the presence of a black hole. Prior to 1783, the entire scientific community was unaware of the existence of black holes. There were no available data on these galactic objects, and no hypotheses suggesting their presence. Incidentally, black holes are *black* in the sense that they are not *directly* observable to us, neither through our eyes nor with our most advanced instruments, for now! This is due to their immense masses generating such powerful gravitational forces that even light cannot escape from them. This state of collective ignorance can be characterized as *unconscious ignorance*. Then, in 1915, Einstein proposed the theory of relativity. His theory predicted that a massive object, such as a black hole, would cause the bending of space–time, which could serve as an indirect mean to detect the presence of black holes. Thus, after 1783, there existed a hypothesis, and later a theory, suggesting how one could potentially validate such a hypothesis. However, there were still no available data or a comprehensive model for black holes (those came later in the 1960s and 1990s). The scientific community remained ignorant about black holes, but this was now a state of *conscious ignorance*. Essentially, there was a lack of both data and a model, but there was a hypothesis to explore. To read more about this check the book of S. Firestein cited in Further Reading.

Now that we have recognized the fundamental role of ignorance in science, one might wonder if the knowledge gained through scientific research would fully compensate such ignorance. The answer is: most likely not! Surprisingly, science not only originates from ignorance but also generates even more ignorance. This statement encompasses many other philosophical concepts, further reinforcing my belief in the interconnectedness of science and philosophy. Implicit in this statement are assumptions such as the existence of an independent and observable reality to investigate, the limitation of science to facts and observable phenomena, and the recognition that we cannot comprehensively understand all aspects of reality through the strict confines of the scientific method. Delving deeper into these aspects is beyond the scope of this book and, more importantly, beyond my expertise. Nonetheless, I would like the reader to grasp a clearer understanding of my intended meaning. When we are consciously ignorant about a particular phenomenon, we can apply the scientific method to eventually explain it. This process involves formulating *hypotheses*, planning *observations* and *data* collection, and ultimately constructing an explanation in the form of a *model* or *theory.* For a model to be considered valid, it must not only account for existing data but also make *predictions* for new phenomena or additional data to be observed. It is through the verification of these predictions that a model is validated. As you can see, science begins with ignorance, and in the pursuit of knowledge, it produces more ignorance in the form of new predictions and data to be collected and explained. In a concise statement, "*science progresses by embracing and expanding ignorance*".

Having established that science emerges from ignorance and progresses by embracing it, we find ourselves confronted with another fundamental question: How should we appropriately pursue knowledge, or, in other words, how should we effectively cultivate ignorance? This question has been the subject of a profound and enduring philosophical debate spanning from the time of Aristotle to the present days. Eminent thinkers such as Francis Bacon, Thomas Aquinas, William of Ockham, Isaac Newton, David Hume, Immanuel Kant, and Karl Popper have expressed their opinions on this matter. The existence of a scientific method, its specific steps, and its significance have been extensively discussed over centuries. Debates about its uniqueness and utility have been raised and explored. However, the focus here is not on engaging in such discussions, but rather on understanding the basic framework that most scientists acknowledge as *the scientific method.*

Firstly, the scientific method is a process aimed at establishing facts with (scientific) rigour. The number of steps in which the scientific method is divided varies in literature from author to author, therefore I decided to enumerate below the formulation that better reflects the practice of my own laboratory:

1. Formulation of a question
2. Formulation of a hypothesis
3. Formulation of a model

4. Derivation of testable predictions
5. Validation (move to 6) or rejection (back to 2)
6. Derivation of new questions from the model (back to 1)

Note then that the scientific method is effectively an infinite loop. It all starts from a question (arising from ignorance) which we answer only to produce a new question and therefore more ignorance. This iterative approach is fundamental to science and is underpinned by the assumption that *we will never come to know everything.*

The first step, *the formulation of a question*, can essentially configure as "there are data that need an understanding" or "we understand a phenomenon well enough to derive new predictions and formulate new questions". Data are to be considered as *objective facts* and are gathered through *observations.*

Hypotheses are statements that try to make sense of the observations. They are our current best guesses to answer the original question. Typically, hypotheses are formulated to be consistent with the observations, i.e., without further investigation they sound logical. This condition is not strictly required but satisfying it can save precious time. Looking logical without further investigation does not, however, ensure the validity of a hypothesis since we have not yet tested whether the predictions arising from such hypothesis are also validated. Care must be taken to never elevate a hypothesis to the rank reserved to objective facts, not even when such a hypothesis has been validated through the scientific method. Facts arise from the physical world and, as such, are objectively valid (again I am assuming there is a reality out there for us to investigate it).

Hypotheses are then used to derive a *theory.* A theory contains a comprehensive set of statements that has two very important features: it describes the existing data; and it can predict other phenomena, already observed or not yet observed. This set of statements is built starting from the hypotheses by using Logic and very often, although not always, Mathematics. The collection of statements plus hypotheses is what we call theory. The predictive nature of a theory is crucial as much as its descriptive nature. Why? Because it is the predictions that are then used to validate the theory in the scientific method.

To be useful, the predictions derived from a theory must be *testable.* Testable means that an experiment can be performed, and data can be gathered to either confirm or contradict the predictions. Nothing stop us to take a given cosmological theory and derive from it the prediction that: "there was no matter, no space, nor time before the big bang". However, hitherto, such prediction is not testable since we cannot go back in time nor re-create the big-bang conditions in a laboratory. Untestable predictions have no place in the scientific method as they cannot be validated. Note, however, that untestable predictions (as well as untestable hypotheses) are perfectly allowed in philosophy or in Mathematics where validation does not need to go through consistency with experimental observations but rather through logic and consistency with all other existing Mathematical theorems.

Validation or rejection of a theory is based on satisfactory match of observations against predictions. Essentially, if the predictions are testable then a test (an experiment, an observation) is run and if passed then the theory and the hypotheses at its core are *temporarily* validated. It is done temporarily because we cannot exclude that new data and future observations may be incompatible with the theory. As a matter of fact, this happens continuously in science, even to "very strong" theories.

At this point, one may raise the question: What happens to a theory built upon a set of hypotheses when certain observations no longer align with the theory? This inquiry has sparked considerable philosophical discourse among experts. Some argue that if a theory encounters observations that are incompatible with it, the entire theory has been disproven and should be discarded. On the other hand, others maintain that the theory can be adjusted or refined to accommodate these previously incompatible observations. The first group of philosophers cautions the second, suggesting that modifying a theory in this way introduces bias and risks forcing the theory to conform to desired outcomes. Conversely, the second group warns the first that completely rejecting a theory can lead to wasted time and disregard for valuable accomplishments.

Predictions are also often used to validate a theory with respect to another one, both describing the same phenomenon. Two theories differs if at least one of the hypotheses is different. And ad-hoc prediction can be put to test to validate the one, and reject the other, theory. It could, however, happen that two different theories are both compatible with the available observations or they are both able to describe a certain number of phenomena, although one theory may fail to describe some phenomena which are still well described by the other one; essentially, one theory is more general than the other. What do we do with these two theories? In science we tend to look at the actual phenomenon we are interested in and pick up the theory that either describes it with the least number of hypotheses or the one which allows us to describe the phenomenon and formulate new predictions in the easiest way (i.e. in the most *practical* way). This is the scientific rendering of the Occam's razor, which states: "*entities must not be multiplied beyond what is necessary*", which, in science, can be read as: "*the simplest theory that is compatible with the observations is the one to be selected*". If interested in reading more about the scientific method, check out the book of M. Di Ventra cited in Further Reading.

Lastly, we might wonder about the tools that facilitate scientific reasoning. How does one progress from a hypothesis to a theory, incorporating predictions, observations, and more? The answer lies in the realm of Logic, with two key tools: *inductive reasoning* and *deductive reasoning.*

Inductive reasoning is a way of reasoning that proceeds by generalization of one or more specific statements. Some specific observations are considered, and a general statement is then produced (induced). Note that even if all specific statements are true, this does not ensure that the induced generalization is also true (but rather only probable). An example of inductive reasoning is the following: "The Bingo caller has called an even number first; she has called a second even number; the third number is also even, then all numbers in the bin are even". This example clearly shows how although each specific statement is true, the induced conclusion is untrue.

Deductive reasoning can be considered the opposite of inductive reasoning. In this tool of logic, one starts from a set of *premises* (a statement that can be true or false) of more general nature to draw (deduce) a more specific conclusion. Note that if all the premises are true then the conclusion is also true. The most classical example of deductive reasoning is Aristotle's syllogism (a deductive argument): "All men are mortal" (first premise), "Socrates is a man" (second premise), then "Socrates is mortal" (conclusion).

When considering these two logical constructs within the framework of the scientific method outlined earlier, we can observe how inductive reasoning is particularly relevant during the phase of formulating hypotheses based on initial data or observations. The fact that the induced conclusions may not be definitively true is inconsequential because hypotheses are not treated as established facts. On the other hand, deductive reasoning assumes a central role in the validation of hypotheses, as deductive conclusions cannot be false if the premises (i.e. the hypotheses in this context) are true. It is important to note that I am not suggesting that one form of reasoning is superior to the other, as both play equally important roles within the scientific method. Additionally, I am not implying that induction is exclusively used for hypothesis formulation while deduction is solely employed for validation. In practice, these two tools are often utilized in different ways. For instance, consider the final step of the scientific method outlined earlier, where new hypotheses are deduced from a theory, initiating a new cycle from the beginning.

NOTE

1. These terms were italicized as they have an important role in the discussion of the scientific method.

Further Reading

TO THE INTERESTED READER who likes to acquire a deeper and more structured knowledge of the topics discussed in this book I suggest to read:

R. P. Feynman, *The Feynman Lectures on Physics (Vol. 1–3)*, 1963–1965, Addison-Wesley Pub. Co., Reading, ISBN: 978-0201500647. Available at: www.feynmanlectures.caltech.edu

J. J. Sakurai and J. Napolitano, *Modern Quantum Mechanics* (3rd ed.), 2020, Cambridge University Press, Cambridge, ISBN: 978-1108473224

C. Cohen-Tannoudji, B. Diu and F. Laloë, *Quantum Mechanics*, 1977, John Wiley & Sons, New York, ISBN: 978-0471164333

L. Susskind and A. Friedman, *Quantum Mechanics—The Theoretical Minimum*, 2014, Allen Lane-Penguin Books, London, ISBN: 978-0141977829

H. Goldstein, *Classical Mechanics* (2nd ed.), 1980, Addison-Wesley Pub. Co., Reading, ISBN: 978-0201029185

L. Susskind and G. Hrabovsky, *Classical Mechanics—The Theoretical Minimum*, 2014, Allen Lane-Penguin Books, London, ISBN: 978-0141976228

S. Firestein, *Ignorance: How It Drives Science*, 2012, Oxford University Press, Oxford, ISBN: 978-0199828074

M. Di Ventra, *The Scientific Method: Reflections from a Practitioner*, 2018, Oxford University Press, Oxford, ISBN: 978-0198825623

Index

Made in the USA
Las Vegas, NV
21 September 2026

36c42178-f217-4dbb-8646-238b7bc564daR02